ALGEBRAIC THEORY OF NUMBERS

Pierre Samuel

Translated from the French by
ALLAN J. SILBERGER

DOVER PUBLICATIONS, INC.
Mineola, New York

Copyright

Bibliographical Note

This Dover edition, first published in 2008, is an unabridged republication of the work originally published in 1970 by Hermann, Paris and Houghton Mifflin Company, Boston.

Library of Congress Cataloging-in-Publication Data

Samuel, Pierre, 1921–
[Théorie algébrique des nombres. English]
Algebraic theory of numbers / Pierre Samuel ; translated from the French by Allan J. Silberger.—Dover ed.
p. cm.
Originally published: Paris : Hermann and Boston : Houghton Mifflin Company, 1970.
Includes bibliographical references and index.
ISBN-13: 978-0-486-46666-8
ISBN-10: 0-486-46666-3
1. Algebraic number theory. I. Title.

QA247.S2513 2008
512.7'4—dc22

2007049457

Manufactured in the United States by LSC Communications
46666305 2017
www.doverpublications.com

Contents

Translator's introduction

LE JUGE: Accusé, soyez fidèle à la vérité.
L'ACCUSÉ: Mais, je suis coupable d'avoir commis le crime.

Professor Samuel has written a beautiful book and, in doing so, he has made an outstanding contribution to the teaching of mathematics. The reader of this book needs to know little beyond the content of a first course in abstract algebra. He will find here a very clear account of classical algebraic number theory, including sufficient examples, applications, and exercises. This book will whet the reader's appetite for and provide him with the background he needs to enjoy the lengthier treatises listed in the bibliography.

The translator wishes to acknowledge, with gratitude, support and assistance he has received while translating this book from the following sources: Bowdoin College, the National Science Foundation (grants GP-7139 and GP-9388), and especially Mrs. Jean Hughes who typed the manuscript.

He also wishes to thank various readers of the earlier French edition, especially Mrs. Germaine Revuz and Mr. Alain Bouvier, who have communicated lists of errata. These mistakes have been corrected in the present English edition.

ALLAN J. SILBERGER
Brunswick, Maine, May 1, 1969

Introduction

The theory of numbers, or arithmetic, is often called "the queen of mathematics". The simplicity of its subject matter (the ordinary integers and their generalizations), the elegance and diversity of its methodology, its numerous unsolved problems strongly attract mathematicians of all classes, whether they be beginners, professional number theorists, or specialists in other branches of mathematics. It should not greatly surprise the reader of this book to learn that its author is an algebraic geometer, who has published no research in what may properly be called arithmetic.

Thus, the reader should not expect an especially profound or comprehensive treatment of number theory. Moreover, he will meet only one of many points-of-view from which one may approach number theory, namely the algebraic point-of-view. Aside from an elementary result of Minkowski concerning lattices in $\mathbf{R}^n$, the beautiful and fertile technique of analytic number theory is entirely omitted from this book.

Emphasis upon the algebraic point-of-view seems to me justifiable for several reasons. First of all, the algebraic point-of-view establishes the context in which number theoretic problems have their most natural formulation. This is true of even those problems which concern only the natural numbers. For example, the problem of finding all integer solutions of the Pell-Fermat equation $X^2 - dY^2 = \pm 1$ (d: a square-free integer) involves in an essential way the study of the quadratic field $\mathbf{Q}[\sqrt{d}]$. For the "great" equation of Fermat $X^n + Y^n = Z^n$ the field of nth roots of unity plays an analogous role. In order to represent an integer as the sum of two (respectively, four) squares, it is advantageous to work in the ring of Gaussian integers (respectively, in a suitably chosen quaternion algebra). The law of quadratic reciprocity involves both quadratic fields and roots of unity. Fields more general than the rational numbers and rings more general than the ordinary integers arise quite naturally when one discusses any of the above problems.

Secondly, although the algebraic approach does not lead to a solution of all number theoretic problems, it does, as the reader will see, nonetheless quickly lead to substantial results. Continuing in the direction of this book, one would reach the deep theorems of class field theory.

Thirdly, even those who prefer analytic number theory will agree that the full generality and power of the analytic approach reveals itself only in the context of number fields and simple algebras, not in investigations involving the rational numbers alone. For example, the Dedekind zeta function of a number field possesses properties quite analogous to those of the ordinary Riemann zeta function and it is not much more difficult to establish these properties in complete generality. A similar statement holds for the various L-series.

Finally, algebraic number theory provides the student with numerous illustrative examples of notions he has encountered in his algebra courses: groups, rings, fields, ideals, quotient rings and quotient fields, homomorphisms and isomorphisms, modules and vector spaces. A further benefit to the student lies in the fact that, in studying algebraic number theory, he will meet many new algebraic notions, notions which are fundamental not only for arithmetic but for other branches of mathematics as well, in particular algebraic geometry. Here are some examples: integrality, field extensions, Galois theory, modules over principal ideal rings, Noetherian rings and modules, Dedekind rings, and rings of fractions.

The preceding describes implicitly what the reader will find in this book and what he will not. I have assumed that the reader knows basic algebra—elementary facts concerning groups, rings, polynomial rings, and vector spaces—the basic facts concerning subobjects, quotient objects, and product objects in these categories—how to pass to the quotient by an ideal or submodule—the various different notions of homomorphism and isomorphism. The reader will find all this (and much more than he needs) in "modern" algebra textbooks, for example the excellent books by R. Godement (Houghton–Mifflin) and S. Lang (Addison–Wesley). This book uses throughout the language and results of "modern" algebra. For the convenience of the reader I have included discussions of the following notions: integrality, algebraic extensions of fields, Galois theory, Noetherian rings and modules, and rings of fractions. I have tried to give short and clear accounts of these topics but without unnecessary sophistication.

This book exists as a consequence of a course in "Higher Mathematics" taught at the University of Paris in 1965 and repeated in 1966. Course notes taken by Alfred Vidal-Madjar, a student at the École Normale Supérieure, to whom I wish gratefully to acknowledge my appreciation, have served as a first draft. Certain parts have come from courses taught at the École Normale Supérieure de Jeunes Filles and at the University of Clermont-sur-Tiretaine. Finally the influence and advice of several mathematicians have been valuable to me. Among them I want particularly to thank the master of my generation, N. Bourbaki, who has had the kindness to show me his unpublished manuscripts, also my friends Emil Grosswald, Georges Poitou, Jean-Pierre Serre, and John Tate.

Bourg-la-Reine, November 1966

A NICOLE

qui a su créer autour de moi
une atmosphère favorable à ce livre

Notations, definitions, and prerequisites

We employ the usual notations from set theory: $\in$, $\subset$, $\cup$, $\cap$. The complement of a subset B of a set A is denoted A − B. ϕ denotes the empty set. The cardinality (or power, or number of elements) of a set A is written card(A); if A is a group, we speak of the order of A.

We assume that the reader is acquainted with the notions of group, ring, field and vector space, as well as with the elementary theory of vector spaces (also called "linear algebra"). In this book, with the exception of Chapter V, § 7, "ring" (respectively, "field") means *commutative* ring (respectively, field) *with unit element.*

Given a finite group G and a subgroup H of G, we recall that card(H) divides card(G). The quotient card(G)/card(H) is called the index of H in G and is denoted (G : H).

Given two subsets A and B of an additive group G, we write A + B for the set of sums $a + b$, $a \in A$, $b \in B$.

Given a ring A, we write A[X] or A[Y] (capital letter) for the ring of polynomials in one variable over A; we write $A[X_1, \ldots, X_n]$ for the polynomials in n variables and A[[X]] for the ring of formal power series.

By convention, a subring A of a ring B contains the unit element of B. Given a ring B, a subring A of B, and an element $x \in B$, we write $A[x]$ for the subring of B generated by A and x, i.e. for the intersection of all subrings of B which contain A and x; it is the set of all sums of the form $a_0 + a_1x + \cdots + a_nx^n (a_i \in A)$. We write $A[x_1, \ldots, x_n]$ for the subring of B generated by A and a finite set $(x_1, \ldots, x_n)$ of elements of B.

A ring A is called an *integral domain* if A contains more than one element and if the product of any two non-zero elements of A is not zero.

An ideal $\mathfrak{b}$ of a ring A is a subgroup of the additive group of A such that $x \in \mathfrak{b}$ and $a \in A$ implies $ax \in \mathfrak{b}$. The whole ring and the set consisting of the element 0 alone (and denoted (0)) are ideals, sometimes called "trivial" ideals. A field has no non-trivial ideals and this fact distinguishes fields from other rings. Given a set of elements (b_i) from a ring A, the intersection of all ideals of A containing the b_i's is an ideal of A, called the ideal generated by the b_i's; it is the set of all elements of the form $\sum_i a_ib_i$ with $a_i \in A$. An ideal generated by a single element b is called principal; notation: Ab or (b).

Let A be a ring and $\mathfrak{b}$ an ideal of A. The equivalence classes $a + \mathfrak{b} (a \in A)$ form a ring, called the quotient ring of A by $\mathfrak{b}$ and denoted $A/\mathfrak{b}$. The ideals of $A/\mathfrak{b}$ are of the form $\mathfrak{b}'/\mathfrak{b}$ where $\mathfrak{b}'$ runs through the set of ideals of A which contain $\mathfrak{b}$. In order that $A/\mathfrak{b}$ be a field it is necessary and sufficient that $\mathfrak{b}$ be maximal among the ideals of A distinct from A. We say then that $\mathfrak{b}$ is a maximal ideal. An ideal $\mathfrak{p}$ is called prime if $A/\mathfrak{p}$ is an integral domain.

Let A and A′ be rings with unit elements e and e'. A homomorphism $f: A \to A'$ is a mapping f of A to A′ such that:

$$f(a + b) = f(a) + f(b), f(ab) = f(a)f(b), \text{ and } f(e) = e'.$$

Let A be a ring. An A-algebra is a pair consisting of a ring B and a homomorphism $\varphi: A \to B$. If A is a field, φ is injective, and we often identify A with its image $\varphi(A)$ (which is a subring of B).

Given a field L and a subfield K of L, we call L an extension of K.

Usually the unit element of a ring A will be denoted by 1.

The notion of *module* over a ring A (or A-module) is the direct generalization of the notion of vector space over a field. An A-module M is an abelian group (the operation is addition) provided with a mapping $A \times M \to M$ (written as multiplication) such that $a(x + y) = ax + ay$, $(a + b)x = ax + bx$, $a(bx) = (ab)x$, $1x = x(a, b \in A, x, y \in M)$. There are notions of submodule and of quotient module. Given two A-modules, M and M′, a homomorphism (or A-linear mapping) of M to M′ is a mapping $f: M \to M'$ such that

$$f(x + y) = f(x) + f(y) \text{ and } f(ax) = af(x) (a \in A, x, y \in M).$$

Given a homomorphism $f: X \to X'$ (of groups, rings, or modules), we call the *kernel* of f and write $\ker(f)$ for the inverse image under f of the zero or identity element of X′ It is a normal subgroup (or an ideal, or a submodule) of X. In order that f be injective it is necessory and sufficient that $\ker(f)$ consist of only the zero or identity element of X. We call the *image* of f the subset $f(X)$ of X′; it is a subgroup (or subring, or submodule) of X′.

Let X and X′ be sets. A mapping f of X to X′ is frequently denoted $f: X \to X'$. When a mapping $f: X \to X'$ is described by the value it takes at an arbitrary element $x \in X$, we use the notation $x \mapsto f(x)$. Thus the sine function, $\sin: \mathbf{R} \to \mathbf{R}$, can be defined by

$$x \mapsto \sum_{n=0}^{\infty} (-1)^n \frac{x^{2n+1}}{(2n+1)!}.$$

We shall employ the usual notations for the following mathematical objects:

N: set of natural numbers (0, 1, 2, ..., n, ...)
(N for "numbers").

Z: ring of rational integers (natural numbers and their negatives)
(Z for "Zahlen").

Q: field of rational numbers (quotients of elements of **Z**)
(Q for "quotients").

R: field of real numbers
(R for "reals").

C: field of complex numbers
(C for "complexes").

$\mathbf{F}_q$: finite field with q elements
(F for "finite" or "field").

Chapter I

Principal ideal rings

1.1 Divisibility in principal ideal rings

Let A be an integral domain, K its field of fractions, x and y elements of K. We shall say that x *divides* y if there exists $a \in A$ such that $y = ax$. Equivalently, we say x is a divisor of y, y is a multiple of x; notation $x \mid y$. This relation between the elements of K depends in an essential manner on the ring A; if there is any confusion possible, we speak of divisibility in K *with respect to* A.

Given $x \in K$ we write Ax for the set of multiples of x. Thus we may write $y \in Ax$ in place of $x \mid y$, or $Ay \subset Ax$. The set Ax is called a principal fractional ideal of K with respect to A; if $x \in A$, Ax is the (ordinary) principal ideal of A generated by x. As the relation of divisibility, $x \mid y$, is equivalent to the *order* relation $Ay \subset Ax$, divisibility possesses the following two properties associated with order relations.

$$(1) \qquad x \mid x; \text{ if } x \mid y \text{ and } y \mid z, \text{ then } x \mid z.$$

On the other hand, if $x \mid y$ and $y \mid x$, one cannot in general conclude that $x = y$; one has only that $Ax = Ay$, which (if $y \neq 0$) means that the quotient xy^{-1} is an invertible element of A; in this case x and y are called *associates*; they are indistinguishable from the point of view of divisibility.

Example. The elements of K which are associates of 1 are the elements invertible in A; they are often called the units in A; they form a group under multiplication, and we shall denote this group A^*. The determination of the units in a ring A is an interesting problem which we shall treat in the case where A is the ring of integers in a number field (see Chapter IV). Here are some simple examples:

(a) If A is a field, $A^* = A - (0)$.
(b) If $A = \mathbf{Z}$, $A^* = \{+1, -1\}$.
(c) The units in the ring of polynomials $B = A[X_1, \ldots, X_n]$, A an integral domain, are the invertible constants; in other words $B^* = A^*$.
(d) The units in the ring of formal power series $A[[X_1, \ldots, X_n]]$ are the formal power series whose constant term is invertible.

Definition 1. A ring A *is called a principal ideal ring if it is an integral domain and if every ideal of* A *is principal.*

We know that the ring **Z** of rational integers is a principal ideal ring. (Recall that any ideal $\mathfrak{a} \neq (0)$ of **Z** contains a least integer $b > 0$. Dividing $x \in \mathfrak{a}$ by b and using the fact that **Z** is Euclidean, one sees that x is a multiple of b.) If K is a field we know that the ring K[X] of polynomials in one variable over K is a principal ideal ring (same method of proof: take a non-zero polynomial $b(X)$ of lowest degree in the given ideal $\mathfrak{a} \neq (0)$ and make use of the fact that K[X] is a Euclidean ring, i.e. the remainder under division of an arbitrary element of $\mathfrak{a}$ by $b(X)$ must be of lower degree than $b(X)$ or zero, which implies zero). This general method shows that any "Euclidean ring" (see [1], Chapter VIII, § 1, exercises, or [9] Chapter 1) is a principal ideal ring. If K is a field, it is easy to see that any non-zero ideal in the ring of formal power series $A = K[[X]]$ is of the form AX^n with $n \geqslant 0$, so that $A = K[[X]]$ is a principal ideal ring too.

Divisibility in the field of fractions K of a principal ideal ring A is particularly simple. We shall review it briefly.

I. Two arbitrary elements u, v of K have a *greatest common divisor* (gcd), i.e. an element d for which the relations

$$\text{(1)} \qquad \text{"}x \mid u \text{ and } x \mid v\text{" and "}x \mid d\text{"}$$

are equivalent. This means the same thing as the assertion that Au and Av have a least upper bound in the partially ordered set of fractional ideals. This least upper bound is $Au + Av$, which is itself a principal fractional ideal, since the ring A is a principal ideal ring (all this is clear for $u, v \in A$; one may reduce to this case by multiplying u and v by a common denominator). We obtain more than the existence of a gcd ("the identity of Bezout"): there exist elements $a, b \in A$ such that the gcd of u and v may be written in the form

$$\text{(2)} \qquad d = au + bv.$$

The greatest common divisor of u and v is obviously unique up to multiplication by units in A.

II. Two arbitrary elements $u, v \in K$ have a *least common multiple* (lcm), i.e. there is an element $m \in K$ for which the relations

$$\text{(3)} \qquad \text{"}u \mid x \text{ and } v \mid x\text{" and "}m \mid x\text{"}$$

are equivalent. This one can see from the fact that sending $t \mapsto t^{-1}$ in K reverses divisibility. So the proof of the existence of least common multiples reduces to the proof of the existence of greatest common divisors. From the relation

$$\text{(4)} \qquad \mathrm{lcm}(u, v) = \mathrm{gcd}(u^{-1}, v^{-1})^{-1} \quad (u, v \neq 0)$$

we easily obtain the following well-known formula

$$\text{(5)} \qquad \mathrm{gcd}(u, v) \cdot \mathrm{lcm}(u, v) = uv.$$

We may also proceed as in (I) to remark that the existence of $\mathrm{lcm}(u, v)$ is equivalent to the existence of a greatest lower bound for Au and Av in the partially ordered set of principal fractional ideals. The greatest lower bound for Au and Av is $Au \cap Av$.

III. Two elements a, b of A are called *relatively prime* if $\gcd(a, b) = 1$. Let us recall the important LEMMA OF EUCLID. *Let a, b, c be elements in a principal ideal ring* A. *If a divides bc and is relatively prime to b, then a divides c.*

A quick proof of Euclid's lemma. By Bezout (2) there exist a' and $b' \in A$ such that $1 = a'a + b'b$; whence $c = a'ac + b'bc$. Since a divides both terms on the right-hand side, a divides c as well.

IV. Finally there is unique factorization into products of primes.

Theorem. Let A *be a principal ideal ring and let* K *be its field of fractions. There exists a subset* $P \subset A$ *such that any* $x \in K$ *may be uniquely expressed in the form*

$$x = u \prod_{p \in P} p^{v_p(x)}, \tag{6}$$

where u is a unit in A *and where the exponents $v_p(x)$ are elements of* $\mathbf{Z}$, *all zero except for a finite subset among them.*

For a more systematic exposition of these topics we refer the reader to [1], *Algèbre*, Chapter VI, § I and Chapter VII, § 1. Part of the theory (more precisely everything which doesn't depend upon Bezout's identity) extends to more general rings than principal ideal rings. We are referring to *unique factorization domains* or *factorial rings*. See [1] or [2] *Algèbre commutative*, Chapter VII, § 3.

1.2 An example: the diophantine equations $X^2 + Y^2 = Z^2$ and $X^4 + Y^4 = Z^4$.

One of the most attractive parts of number theory is the study of *diophantine equations*. One considers polynomial equations $P(X_1, \ldots, X_n) = 0$ with coefficients in $\mathbf{Z}$ (respectively, in $\mathbf{Q}$) and one seeks solutions (x_i) in $\mathbf{Z}$ (respectively, in $\mathbf{Q}$). One can replace $\mathbf{Z}$ (respectively, $\mathbf{Q}$) by more general rings A (respectively, fields K). We will give an example later (§ 6).

We are going to study here two special cases of Fermat's famous equation:

$$X^n + Y^n = Z^n. \tag{1}$$

Fermat claimed to have shown that, for $n \geq 3$, this equation has no non-trivial integer solution (x, y, z). His proof has never been found. Numerous mathematicians have since Fermat's time worked intensively on this problem. They have shown that Fermat's claim is true for a great many values of the exponent n. Nevertheless, no general proof (i.e. valid for any n) has been found.

Present-day opinion holds that, in his "proof", Fermat made a mistake, but a mistake worthy of a first-class mathematician. For example he might have conceived the idea (ingenious for his time) of working in the ring of integers of a field containing nth roots of unity; he may have believed that such a ring is always a principal ideal ring. In fact, we know how to prove Fermat's claim for any exponent n for which the ring of nth roots of unity is a principal ideal ring. However, this is not the case for all n. For n prime, this ring is a principal ideal ring only for finitely many values of n.[1]

For $n = 2$, equation (1) has integer solutions, e.g. (3, 4, 5). One can give a complete description of all the integer solutions of (1).

1. Cf. C. L. Siegel "Gesammelte Werke", Part III, pp. 436–442.

Theorem 1. If x, y, z are positive integers such that $x^2 + y^2 = z^2$, then there exists an integer d and two relatively prime integers u and v such that (except, possibly, for a permutation of x and y):

$$(2) \qquad x = d(u^2 - v^2), \quad y = 2duv, \quad \text{and} \quad z = d(u^2 + v^2).$$

Proof. An easy calculation shows that formula (2) gives solutions for $X^2 + Y^2 = Z^2$. Conversely, let x, y, and z be positive integers such that $x^2 + y^2 = z^2$. After dividing x, y, z by their greatest common divisor, we may assume that the three numbers are relatively prime. It follows that they are pairwise relatively prime as well; for example, if x and z have a common prime factor p, then p divides $y^2 = z^2 - x^2$ and, therefore, also y. In particular, two of the numbers x, y, z are odd; the third is necessarily even. The numbers x and y cannot both be odd, for, if they were, we would have $x^2 \equiv 1 \pmod 4$, $y^2 \equiv 1 \pmod 4$, and $z^2 \equiv 2 \pmod 4$, which contradicts the fact that z^2 is a square. We have, then, after possibly switching x and y:

$$(3) \qquad x \text{ odd}, y \text{ even, and } z \text{ odd}.$$

Note that

$$(4) \qquad y^2 = z^2 - x^2 = (z - x)(z + x).$$

Since the greatest common divisor of $2x$ and $2z$ is 2, and since $2x = (z + x) - (z - x)$, $2z = (z + x) + (z - x)$, the greatest common divisor of $z - x$ and $z + x$ can only be 2. Put $y = 2y'$, $z + x = 2x'$, $z - x = 2z'$, where y', x', and z' are integers (since y, $z + x$, and $z - x$ are even by (3)). We have $y'^2 = x'z'$. Since x' and z' are relatively prime, we see that x' and z' are squares u^2 and v^2; in fact any prime factor of y'^2 appears, with an even exponent, either in the prime factorization of x' or in that of z', but not in both. We thus have $z + x = 2u^2$, $z - x = 2v^2$, and $y^2 = 2u^2 \cdot 2v^2$; whence, $x = u^2 - v^2$, $y = 2uv$, and $z = u^2 + v^2$. Here u and v are relatively prime, since otherwise x, y, z would have a common prime factor. Multiplying through by the greatest common divisor of x, y, z, call it d, we obtain (2). Q.E.D.

Theorem 2. The equation $X^4 + Y^4 = Z^2$ has no solution in positive integers x, y, z.

Proof. If there is a solution (x, y, z), where x, y, and z are positive integers, then there is such a solution in which z is *minimal*. In this case, x, y, and z are pairwise relatively prime; if for example x and y have a common prime factor p, then p^4 divides z^2, so p^2 divides z and $(x/p, y/p, z/p^2)$ would be a solution, contradicting the minimality of z. The two other cases are analogous and even easier. As our equation may be written as $(X^2)^2 + (Y^2)^2 = Z^2$, we may apply Theorem 1 to it. After possibly permuting x and y we see that there are positive relatively prime integers u and v such that

$$(5) \qquad x^2 = u^2 - v^2, \quad y^2 = 2uv, \quad \text{and} \quad z = u^2 + v^2.$$

Since $4 \mid y^2$, the relation $y^2 = 2uv$ implies that one of the two numbers u and v is even; the other is necessarily odd. Thus, u even and v odd entails $u^2 \equiv 0 \pmod 4$ and $v^2 \equiv 1 \pmod 4$, whence $x^2 = u^2 - v^2 \equiv -1 \pmod 4$, an impossibility. So u is odd and $v = 2v'$. The relation $y^2 = 4uv'$ and the fact that u and v' are relatively prime implies that u and v' are squares a^2 and b^2. We apply Theorem 1 once more, this time to the equation $x^2 + v^2 = u^2$ (cf. (5)). Since x and u are odd, v even, and x, v, and u pairwise prime, we obtain two relatively prime positive integers c and d such that:

(6) $$x = c^2 - d^2, \quad v = 2cd, \quad \text{and} \quad u = c^2 + d^2.$$

Now, from $v = 2v' = 2b^2$, it follows that $cd = b^2$, so that c and d are again squares x'^2 and y'^2 (they are relatively prime). Since $u = a^2$, (6) may be rewritten as

(7) $$a^2 = x'^4 + y'^4,$$

which is of the same form as the original equation. On the other hand, by (5), $z = u^2 + v^2 = a^4 + 4b^4 > a^4$, whence $z > a$, which contradicts the minimality of z. Theorem 2 is proved.

A slight variant of our proof shows that, given a solution (x, y, z) in positive integers of $X^4 + Y^4 = Z^2$, one may construct an infinite sequence (x_n, y_n, z_n) of such solutions, where the sequence (z_n) is strictly decreasing. This is an absurdity. This method of proof is called the method of infinite descent and is due to Fermat.

Corollary. The equation $X^4 + Y^4 = Z^4$ *has no positive integer solutions.*

Proof. This equation may be written in the form $X^4 + Y^4 = (Z^2)^2$, to which Theorem 2 applies.

1.3. Some lemmas concerning ideals; Euler's φ-function

Let $n \geq 1$ be a natural number. We write $\varphi(n)$ for the number of integers q, $0 \leq q \leq n$, such that q and n are relatively prime (since 0 and n are divisible by n, it is equivalent to take $1 \leq q \leq n - 1$ for any $n > 1$; set $\varphi(1) = 1$). The function φ so defined is called Euler's φ-function. If p is a prime number, then clearly:

(1) $$\varphi(p) = p - 1.$$

For $n = p^s$, a power of a prime, the integers relatively prime to n are those integers which are not multiples of p. There are p^{s-1} multiples of p between 1 and p^s. Therefore,

(2) $$\varphi(p^s) = p^s - p^{s-1} = p^{s-1}(p-1).$$

Now we intend to calculate $\varphi(n)$ by making use of the fact that n may be expressed as a product of powers of primes. For this purpose we need some properties of $\varphi(n)$ and we need some lemmas concerning ideals. These lemmas will be useful later.

Proposition 1. Let $n \geq 1$ *be a natural number. The value* $\varphi(n)$ *of Euler's* φ*-function equals the number of elements of* $\mathbf{Z}/n\mathbf{Z}$ *which generate this group. It also equals the number of units in the ring* $\mathbf{Z}/n\mathbf{Z}$.

Proof. Let us recall that each congruence class mod $n\mathbf{Z}$ contains a unique integer q such that $0 \leq q \leq n - 1$. For such an integer q we write $\bar{q}$ for its residue class mod $n\mathbf{Z}$. It suffices to prove the following implications: q relatively prime to $n \Rightarrow \bar{q}$ a unit in the ring $\mathbf{Z}/n\mathbf{Z} \Rightarrow \bar{q}$ generates the additive group $\mathbf{Z}/n\mathbf{Z} \Rightarrow q$ relatively prime to n. If q is relatively prime to n, Bezout's identity (§ 1, (2)) implies that there are integers x and y

such that $qx + ny = 1$; whence $\bar{q}\cdot\bar{x} = \bar{1}$, so $\bar{q}$ is a unit in $\mathbf{Z}/n\mathbf{Z}$. Let $\bar{q}$ be a unit in $\mathbf{Z}/n\mathbf{Z}$. Writing x for an integer such that $\bar{q}\cdot\bar{x} = 1$, we see that $\bar{a} = \bar{a}\cdot\bar{x}\cdot\bar{q}$ (in the ring $\mathbf{Z}/n\mathbf{Z}$), where $\bar{a}$ is an arbitrary element of $\mathbf{Z}/n\mathbf{Z}$ $(0 \leq a < n)$. It follows that $\bar{a} = (ax)\cdot\bar{q}$ (in the additive group $\mathbf{Z}/n\mathbf{Z}$), so $\bar{q}$ generates the group $\mathbf{Z}/n\mathbf{Z}$. Finally, if $\bar{q}$ generates $\mathbf{Z}/n\mathbf{Z}$, there is an x such that $x\cdot\bar{q} = \bar{1}$, thus such that $xq \equiv 1 (\text{mod } n)$; thus there exists an integer y for which $xq - 1 = yn$, so $1 = xq - yn$. This is an instance of Bezout's identity, which shows that q is relatively prime to n. Q.E.D.

Lemma 1. Let A *be a ring,* $\mathfrak{a}$ *and* $\mathfrak{b}$ *ideals of* A *such that* $\mathfrak{a} + \mathfrak{b} = A$. *Then* $\mathfrak{a} \cap \mathfrak{b} = \mathfrak{a}\mathfrak{b}$ *and the canonical homomorphism* $\varphi: A \to A/\mathfrak{a} \times A/\mathfrak{b}$ *induces an isomorphism* $[\theta: A/\mathfrak{a}\mathfrak{b} \to A/\mathfrak{a} \times A/\mathfrak{b}.]$

Recall that the homomorphism φ sends any $x \in A$ into the pair consisting of the class of x mod $\mathfrak{a}$ and the class of x mod $\mathfrak{b}$.

Proof. We know that, in general, $\mathfrak{a}\mathfrak{b} \subset \mathfrak{a}$ and $\mathfrak{a}\mathfrak{b} \subset \mathfrak{b}$, so $\mathfrak{a}\mathfrak{b} \subset \mathfrak{a} \cap \mathfrak{b}$. Let $x \in \mathfrak{a} \cap \mathfrak{b}$. Since $\mathfrak{a} + \mathfrak{b} = A$, there are elements $a \in \mathfrak{a}$ and $b \in \mathfrak{b}$ such that $a + b = 1$. Thus $x = ax + xb$ is a sum of two elements of $\mathfrak{a}\mathfrak{b}$, whence $x \in \mathfrak{a}\mathfrak{b}$ and $\mathfrak{a} \cap \mathfrak{b} \subset \mathfrak{a}\mathfrak{b}$. Therefore $\mathfrak{a}\mathfrak{b} = \mathfrak{a} \cap \mathfrak{b}$.

Clearly $\mathfrak{a} \cap \mathfrak{b}$ is the kernel of φ. Since $\mathfrak{a} \cap \mathfrak{b} = \mathfrak{a}\mathfrak{b}$, φ is constant on each residue class mod $\mathfrak{a}\mathfrak{b}$. Thus, we obtain a mapping $\theta: A/\mathfrak{a}\mathfrak{b} \to A/\mathfrak{a} \times A/\mathfrak{b}$, which is obviously a homomorphism. Since $\varphi^{-1}(0) = \mathfrak{a}\mathfrak{b}$, $\theta^{-1}(0) = (0)$, so θ is injective. It remains to show that θ is surjective.

We have "passed to the quotient ring" in the course of the above argument. Henceforth, we shall be much more brief in explaining analogous arguments.

In order to show that θ (or, equivalently, φ) is surjective, we have to find, for any pair $y \in A$ and $z \in A$, an element $x \in A$ such that $x + \mathfrak{a} = y + \mathfrak{a}$ and $x + \mathfrak{b} = z + \mathfrak{b}$. Take $a \in \mathfrak{a}$ and $b \in \mathfrak{b}$ such that $a + b = 1$. Set $x = az + by$. Modulo $\mathfrak{a}$, $x \equiv by \equiv (1 - a)y \equiv y - ay \equiv y$; similarly, $x \equiv z$ mod $\mathfrak{b}$. Q.E.D.

Lemma 2. Let A *be a ring and* $(\mathfrak{a}_i)_{1 \leq i \leq r}$ *a finite set of ideals of* A *such that* $\mathfrak{a}_i + \mathfrak{a}_j = A$ *for* $i \neq j$. *Then there is a canonical isomorphism of* $A/\mathfrak{a}_1 \ldots \mathfrak{a}_r$ *onto* $\prod_{i=1}^{r} A/\mathfrak{a}_i$.

Proof. Lemma 1 is the case $r = 2$ of Lemma 2. We proceed by induction on r. Put $\mathfrak{b} = \mathfrak{a}_2 \ldots \mathfrak{a}_r$. Let us show that $\mathfrak{a}_1 + \mathfrak{b} = A$. For $i \geq 2$ we have $\mathfrak{a}_1 + \mathfrak{a}_i = A$, so there are elements $c_i \in \mathfrak{a}_1$ and $a_i \in \mathfrak{a}_i$ such that

$$c_i + a_i = 1, \quad 1 = \prod_{i=1}^{r} (c_i + a_i) = c + a_2 \ldots a_n,$$

where c is a sum of terms each of which contains at least one c_i as a factor. Therefore, $c \in \mathfrak{a}_1$. As $a_2 \ldots a_r \in \mathfrak{b}$, it follows that $\mathfrak{a}_1 + \mathfrak{b} = A$.

By Lemma 1, $A/\mathfrak{a}_1\mathfrak{b}$ is isomorphic to $A/\mathfrak{a}_1 \times A/\mathfrak{b}$. According to the induction hypothesis $A/\mathfrak{b} = A/\mathfrak{a}_2 \ldots \mathfrak{a}_r$ which is isomorphic to $A/\mathfrak{a}_2 \times \cdots \times A/\mathfrak{a}_r$. The lemma follows by composing these isomorphisms. Q.E.D.

Let us apply these lemmas to $\mathbf{Z}$.

Proposition 2. Let n *and* n' *be relatively prime integers. Then the ring* $\mathbf{Z}/nn'\mathbf{Z}$ *is isomorphic to the product ring* $\mathbf{Z}/n\mathbf{Z} \times \mathbf{Z}/n'\mathbf{Z}$.

Proof. This is a special case of Lemma 1, the hypothesis $n\mathbf{Z} + n'\mathbf{Z} = \mathbf{Z}$ being Bezout's identity.

Corollary 1. If n and n' are relatively prime positive integers, then $\varphi(nn') = \varphi(n)\varphi(n')$.

Proof. $\varphi(nn')$ is the number of units in $\mathbf{Z}/nn'\mathbf{Z}$ (Proposition 1), which is isomorphic to $\mathbf{Z}/n\mathbf{Z} \times \mathbf{Z}/n'\mathbf{Z}$. Now an element (α, β) of a ring product is invertible if and only if each of its components α, β is invertible. Thus our assertion follows from Proposition 1.

Corollary 2. Let n be a positive integer and let $n = p_1^{\alpha_1} \ldots p_r^{\alpha_r}$ *be its prime factorization. Then* $\varphi(n) = n(1 - 1/p_1) \cdots (1 - 1/p_r)$.

Proof. By Corollary 1 $\varphi(n) = \varphi(p_1^{\alpha_1}) \ldots \varphi(p_r^{\alpha_r})$. By (2) $\varphi(p_i^{\alpha_i}) = p_i^{\alpha_i - 1}(p_i - 1) = p_i^{\alpha_i}(1 - 1/p_i)$. Multiplication gives our formula.

1.4. Some preliminaries concerning modules

Before studying modules over a principal ideal ring, we make some remarks concerning modules over arbitrary commutative rings.

Let A be a commutative ring and let I be a set. Let $A^{(I)}$ denote the set of sequences $(a_i)_{i \in I}$, indexed by I, of elements of A *such that* $a_i = 0$ except for a *finite* number of indices $i \in I$. Thus $A^{(I)}$ is a subset of the cartesian product set A^I, and also a submodule of A^I—if one provides A^I with the A-module structure defined by componentwise addition and scalar multiplication.

If I is finite, then $A^{(I)} = A^I$.

For $j \in I$, the sequence $(\delta_{ji})_{i \in I}$ such that $\delta_{jj} = 1$ and $\delta_{ji} = 0$ for $i \neq j$ is an element e_j of $A^{(I)}$. Every element $(a_j)_{j \in I}$ of $A^{(I)}$ has a unique expression as a (finite) linear combination of the e_j. More precisely,

$$(a_j)_{j \in I} = \sum_{j \in I} a_j e_j \tag{1}$$

(note that, in the summation on the right, all the terms are zero except for a finite number, so that the summation makes sense).

We call $(e_j)_{j \in I}$ the *canonical base* for $A^{(I)}$.

Let A be a ring, M an A-module, and $(x_i)_{i \in I}$ a family of elements of M. To every element $(a_i)_{i \in I}$ of $A^{(I)}$ let us associate the element $\sum_i a_i x_i$ of M (as before, the summation makes sense). Thus we obtain a mapping $\varphi : A^{(I)} \to M$, which is obviously linear. If $(e_i)_{i \in I}$ is the canonical base for $A^{(I)}$, then $\varphi(e_i) = x_i$ for any $i \in I$. The equivalence of the following statements is immediate:

(2) $(x_i)_{i \in I}$ is a linearly independent set if and only if φ is injective.

(3) $(x_i)_{i \in I}$ generates M if and only if φ is surjective. If φ is *bijective* $(x_i)_{i \in I}$ is called a *base* for M. This means that every element of M has a unique expression as a linear combination of the elements $(x_i)_{i \in I}$. A module M which has a base is called a *free module.*

In contrast to the case of vector spaces over a field, a module over a ring does not necessarily have a base. For example, the $\mathbf{Z}$-module $\mathbf{Z}/n\mathbf{Z}$ for $n \neq 0$ or 1. In the ensuing discussion we shall show that certain modules are free. This type of result is seldom trivial.

A module is said to be of *finite type* if it contains a finite generating set. The following theorem is basic for the study of the properties of Noetherian rings and modules. We develop this topic further in Chapter III.

Theorem 1. Let A *be a ring and* M *an* A*-module. The following conditions are equivalent.*

(a) *Every non-empty family of submodules of* M *contains a maximal element (under the relation of inclusion).*

(b) *Every increasing sequence* $(M_n)_{n\geq 0}$ *(again for the relation of inclusion) of submodules of* M *is stationary (i.e. there exists* n_0 *such that* $M_n = M_{n_0}$ *for all* $n \geq n_0$*).*

(c) *Every submodule of* M *is of finite type.*

Proof. Let us show that (a) implies (c). Let E be a submodule of M and let Φ be the collection consisting of all submodules of finite type of E. Then Φ is not empty, since $(0) \in \Phi$. By (a) Φ contains a maximal element F. For $x \in E$, $F + Ax$ is a submodule of finite type of E (it is generated by the union of $\{x\}$ and any finite set of generators for F). Thus $F + Ax = F$, since $F + Ax \supset F$ and F is maximal. Therefore, $x \in F$, $E \subset F$, $E = F$, and E is of finite type.

We prove now that (c) implies (b). Let $(M_n)_{n\geq 0}$ be an increasing sequence of submodules of M. Then $E = \bigcup_{n\geq 0} M_n$ is a submodule of M. By (c) it contains a finite set of generators $(x_1, \ldots, x_q)$. For every i there is an index $n(i)$ such that $x_i \in M_{n(i)}$. Let n_0 be the largest of the $n(i)$'s. Then $x_i \in M_{n_0}$ for all i, so $E \subset M_{n_0}$ and $E = M_{n_0}$. For $n \geq n_0$ the inclusion relations $M_{n_0} \subset M_n \subset E$ and the equality $M_{n_0} = E$ imply that $M_{n_0} = M_n$. Thus the sequence (M_n) is stationary beyond n_0.

It remains to prove that (b) implies (a). The equivalence of (a) and (b) is a special case of the following lemma concerning partially ordered sets.

Lemma 1. Let T *be a partially ordered set. The following statements are equivalent:*

(a) *Every non-empty subset of* T *contains a maximal element.*

(b) *Every increasing sequence* $(t_n)_{n\geq 0}$ *of elements of* T *is stationary.*

Proof. (a) $\Rightarrow$ (b): Let t_q be a maximal element of the increasing sequence (t_n). Then, for $n \geq q$, $t_n \geq t_q$ (the sequence increases), so $t_n = t_q$ (maximality).

(b) $\Rightarrow$ (a): Suppose there exists a subset S of T which does not contain a maximal element. Then, for any $x \in S$, the set of elements of S which are larger than x is non-empty. By the axiom of choice there exists a mapping $f\colon S \to S$ such that $f(x) > x$ for all $x \in S$. Since S is not empty, one may choose $t_0 \in S$ and define by induction the sequence $(t_n)_{n\geq 0}$ by setting $t_{n+1} = f(t_n)$. This sequence is strictly increasing; it is therefore not stationary. This contradicts (b), so (b) $\Rightarrow$ (a) is established.

Corollary of theorem 1. In a principal ideal ring A, *every non-empty family of ideals contains a maximal element.*

Proof. If one considers A as a module over itself, its submodules are its ideals. As all ideals are principal, they are A-modules generated by a single element, thus of finite type. The corollary follows from the implication (c) $\Rightarrow$ (a) of Theorem 1.

1.5. Modules over principal ideal rings

Let A be an integral domain and let K be its field of fractions. A free A-module (isomorphic to an $A^{(I)}$ for some I) may be injected into a vector space over K ($K^{(I)}$ in the case of $A^{(I)}$). It follows that the same thing is true for any submodule M of a free A-module. The dimension of the subspace generated by M is called the rank of M. If M is itself free and admits a base having n elements, then the rank of M is n.

Theorem 1. Let A *be a principal ideal ring,* M *a free* A*-module of rank n, and* M′ *a submodule of* M. *Then:*

(a) M′ *is free of rank* q, $0 \leq q \leq n$.
(b) *If* $M' \neq (0)$, *there exists a base* $(e_1, \ldots, e_n)$ *of* M *and non-zero elements* $a_1, \ldots, a_q \in A$ *such that* $(a_1e_1, \ldots, a_qe_q)$ *is a base of* M′ *and such that* a_i *divides* a_{i+1}, $1 \leq i \leq q-1$.

Proof. The theorem is trivial for $M' = (0)$, so we may assume $M' \neq (0)$. Let L(M, A) be the set of linear forms on M. For $u \in L(M, A)$, $u(M')$ is a submodule of A, an ideal of A. We may write $u(M') = Aa_u$ with $a_u \in A$, since the ideal is principal. Let $u \in L(M, A)$ be such that Aa_u is maximal among the $Aa_v (v \in L(M, A))$ (§ 4, corollary of Theorem 1). Let us take a base $(x_1, \ldots, x_n)$ of M, which identifies M with A^n. Let $pr_i : M \to A$ be the projection on the ith coordinate, i.e. $pr_i(x_j) = \delta_{ij}$. Since $M' \neq (0)$, for at least one i, $1 \leq i \leq n$, $pr_i(M')$ is not (0). Thus $a_u \neq (0)$. By our construction there exists $e' \in M'$ such that $u(e') = a_u$. *Let us show that for every* $v \in L(M, A)$, $a_u \mid v(e')$. Indeed, if $d = \gcd(a_u, v(e'))$, then $d = ba_u + cv(e')$ with $b, c \in A$, whence $d = (bu + cv)(e')$. Since $bu + cv$ is a linear form on M, $Aa_u \subset Ad \subset u(M')$. The maximality of Aa_u implies that $Ad = Aa_u$, so a_u must divide $v(e')$.

In particular, $a_u \mid pr_i(e')$, so let $pr_i(e') = a_u b_i$ with $b_i \in A$. Put $e = \sum_{i=1}^{n} b_i x_i$. Then $e' = a_u e$. Since $u(e') = a_u = a_u \cdot u(e)$, it follows that $u(e) = 1$ (note that $a_u \neq 0$). Let us show that

(1) $M = \mathrm{Ker}(u) + Ae$
and (2) $M' = (M' \cap \mathrm{Ker}(u)) + Ae'$ (where $e' = a_u e$),

the sums being direct.

1. For every $x \in M$, $x = u(x)e + (x - u(x)e)$. We see that $u(x - u(x)e) = u(x) - u(x)u(e) = 0$, since $u(e) = 1$, so $x - u(x)e \in \mathrm{Ker}(u)$. This shows that $Ae + \mathrm{Ker}(u) = M$; obviously $Ae \cap \mathrm{Ker}(u) = (0)$.
2. For $y \in M'$, $u(y) = ba_u$ with $b \in A$, so $y = ba_u e + (y - u(y)e) = be' + (y - u(y)e)$.

   Again it is clear that $y - u(y)e \in \mathrm{Ker}(u)$ and also that $y - u(y)e = y - be' \in M'$, i.e. $y - u(y)e \in M' \cap \mathrm{Ker}(u)$ and $be' \in Ae' \subset Ae$. This entails (2).

Now we prove (a) by induction on the rank q of M′. If $q = 0$, $M' = (0)$ and there is nothing to prove. If $q > 0$, $M' \cap \mathrm{Ker}(u)$ is of rank $q - 1$ according to (2), and is therefore free according to the induction hypothesis. As, in (2), the sum is direct, we obtain a base for M′ by adding e' to a base for $M' \cap \mathrm{Ker}(u)$. Thus M′ is free and (a) is true.

To prove (b) we argue by induction on the rank n of M. Again the case $n = 0$ is trivial. By (a), Ker(u) is free and of rank $n - 1$, since, in (1), the sum is direct. We apply the induction hypothesis to the free module Ker(u) and to its submodule $M' \cap \text{Ker}(u)$: if $M' \cap \text{Ker}(u) \neq (0)$, there exists $q \leq n$, a base $(e_2, \ldots, e_n)$ of Ker (u), and there are non-zero elements $a_2, \ldots, a_q$ of A such that $(a_2e_2, \ldots, a_qe_q)$ is a base for $M' \cap \text{Ker}(u)$ and such that a_i divides a_{i+1}, $2 \leq i \leq q - 1$. Keeping the same notations as above, we set $a_1 = a_u$ and $e_1 = e$. Then $(e_1, e_2, \ldots, e_n)$ is a base for M (according to (1)), and $(a_1e_1, \ldots, a_qe_q)$ is a base for M' (according to (2) and the fact that $e' = a_1e_1$). It remains to prove that $a_1 \mid a_2$. Let v be the linear form on M defined by the relation $v(e_1) = v(e_2) = 1$ and $v(e_i) = 0$ for $i \geq 3$. Then $a_1 = a_u = v(a_ue_1) = v(e') \in v(M')$, so $Aa_u \subset v(M')$. By the maximality of Aa_u we may conclude that $v(M') = Aa_u = Aa_1$. Since $a_2 = v(a_2e_2) \in v(M')$, we see that $a_2 \in Aa_1$, i.e. $a_1 \mid a_2$. Q.E.D.

The ideals Aa_i of Theorem 1 are called the *invariant factors* of M' in M. One can show that they are uniquely determined by M and M' ([1], Chapter VII, § 3).

Corollary 1. Let A *be a principal ideal ring. Let* E *be an* A-*module of finite type. Then* E *is isomorphic to a product* $(A/\mathfrak{a}_1) \times (A/\mathfrak{a}_2) \times \cdots \times (A/\mathfrak{a}_n)$, *where the* $\mathfrak{a}$'s *are ideals of* A *such that* $\mathfrak{a}_1 \supset \mathfrak{a}_2 \supset \cdots \supset \mathfrak{a}_n$.

Proof. Let $(x_1, \ldots, x_n)$ be a finite set of generators of E. According to the beginning of § 4, there is a surjective homomorphism $\varphi: A^n \to E$, such that E is isomorphic to $A^n/\text{Ker}(\varphi)$. By Theorem 1, there is a base $(e_1, \ldots, e_n)$ of A^n, an integer $q \leq n$, and non-zero elements $a_1, \ldots, a_q$ of A such that $(a_1e_1, \ldots, a_qe_q)$ is a base of Ker (φ) and such that a_i divides a_{i+1} for all i, $1 \leq i \leq q - 1$. Put $a_p = 0$ for $q + 1 \leq p \leq n$. Then $A^n/\ker(\varphi)$ is isomorphic to the product of the Ae_i/Aa_ie_i $(1 \leq i \leq n)$, and Ae_i/Aa_ie_i is isomorphic to A/Aa_i. Putting $\mathfrak{a}_i = Aa_i$, we obtain the corollary. Q.E.D.

We shall say that a module E over an integral domain A is *torsion free* if the relation $ax = 0$ $(a \in A, x \in E)$ implies $a = 0$ or $x = 0$.

Corollary 2. Over a principal ideal ring A, *every module* E *of finite type which is torsion free is free.*

Proof. We make use of Corollary 1: $E \simeq (A/\mathfrak{a}_1) \times \cdots \times (A/\mathfrak{a}_n)$. Suppressing the factors which are zero, we may suppose that $\mathfrak{a}_i \neq A$ for all i. If $\mathfrak{a}_1 \neq (0)$, if a is a non-zero element of $\mathfrak{a}_1$, if x_1 is a non-zero element of $A/\mathfrak{a}_1$, and if $x = (x_1, 0, \ldots, 0)$, then $ax = 0$ —contradicting the fact that E is torsion free. Thus $\mathfrak{a}_1 = (0)$, $\mathfrak{a}_i = (0)$ for all i (since $\mathfrak{a}_i \subset \mathfrak{a}_1$), and E is isomorphic to A^n.

The hypothesis that E is of finite type is essential: for example **Q** is a torsion free **Z**-module which is not free.

Corollary 3. Over a principal ideal ring A, *every module* E *of finite type is isomorphic to a finite product of modules* M_i, *where each* M_i *is equal to* A *or to a quotient* A/Ap^s *with* p *prime.*

Proof. We make use of Corollary 1 and we decompose each factor A/Aa, where $a \neq 0$, by means of § 3, Lemma 2: if $a = up_1^{s_1} \ldots p_r^{s_r}$ is the prime factorization of a, A/Aa is isomorphic to the product of the $A/Ap_i^{s_i}$'s.

Corollary 4. Let G *be a finite commutative group. There exists* $x \in G$ *whose order is the least common multiple of the orders of the elements of* G.

Proof. A commutative group is a **Z**-module (the operation being addition). According to Corollary 1,

$$G \simeq \mathbf{Z}/a_1\mathbf{Z} \times \cdots \times \mathbf{Z}/a_n\mathbf{Z},$$

where $a_1 \mid a_2 \mid \ldots \mid a_n$. We have $a_i \neq 0$ for all i; otherwise G would be infinite. We write y for the residue class of 1 in $\mathbf{Z}/a_n\mathbf{Z}$ and we put $x = (0, \ldots, 0, y)$. The order of x is obviously a_n. For $z = (z_1, \ldots, z_n) \in G$, we have $a_n z = 0$, since a_i divides a_n for all i. Therefore a_n is a multiple of the order of z and x is the element sought.

1.6. Roots of unity in a field

Theorem 1. Let K *be a field. Every finite subgroup* G *of the multiplicative group* K* *consists of roots of unity and is cyclic.*

Proof. According to § 5, Corollary 4 of Theorem 1, there exists $z \in G$ whose order n is such that $y^n = 1$ for every $y \in G$. Since a polynomial of degree n over a field (for example $X^n - 1$) has at most n roots in the field, the number of elements of G is at most n. Now, inasmuch as z has order n, G contains the n elements $z, z^2, \ldots, z^n = 1$, which are all distinct. Thus G is comprised of these elements and is cyclic. Q.E.D.

If a field K contains n nth roots of unity, they form a cyclic group of order n (isomorphic to $\mathbf{Z}/n\mathbf{Z}$). A generator of this group is called a primitive nth root of unity; every nth root of unity is thus a power of such a primitive root. According to § 3, Proposition 1, the number of these primitive roots is $\varphi(n)$.

1.7. Finite fields

Let K be a field. There is a unique ring homomorphism $\varphi: \mathbf{Z} \to K$ (defined by $\varphi(n) = 1 + 1 + \cdots + 1$, n times, for $n \geq 0$ and by $\varphi(-n) = -\varphi(n)$). If φ is injective, it identifies **Z** with a subring of K; then K also contains the field of fractions **Q** of **Z**. In this case we say that K is *of characteristic* 0. If φ is not injective, its kernel is an ideal $p\mathbf{Z}$ where $p > 0$; then $\mathbf{Z}/p\mathbf{Z}$ is identified with a subring of K; thus $\mathbf{Z}/p\mathbf{Z}$ is an integral domain (in fact, a field) from which it follows that p is a *prime number*. We say, in this case, that K is *of characteristic* p. From here on, we write $\mathbf{F}_p$ for $\mathbf{Z}/p\mathbf{Z}$.

The subfield, **Q** or $\mathbf{F}_p$, of K is the smallest subfield of K; it is called the prime subfield of K. For every prime number p there exist fields of characteristic p, e.g. $\mathbf{F}_p$.

Proposition 1. If K *is a field of characteristic* $p \neq 0$, *then* $px = 0$ *for every* $x \in K$ *and* $(x + y)^p = x^p + y^p$ *for every* $x, y \in K$.

Proof. For $x \in K$, we have $p \cdot x = (p \cdot 1) \cdot x = 0 \cdot x = 0$. On the other hand, the binomial formula gives

$$(x + y)^p = x^p + y^p + \sum_{j=1}^{p-1} \binom{p}{j} x^j y^{p-j}.$$

The binomial coefficient $\binom{p}{j}$ is an integer; its value is $p!/[j!(p-j)!]$. Inasmuch as the prime p appears in the numerator but not in the denominator, $\binom{p}{j}$ is a multiple of p for $1 \leq j \leq p-1$. The intermediate terms in the expansion of $(x+y)^p$ thus vanish in a field of characteristic p.

By induction one sees that $(x+y)^{p^n} = x^{p^n} + y^{p^n}$ for every $n \geq 0$.

Theorem 1. Let K *be a finite field. Set* $q = \text{card}\,(\text{K})$. *Then:*

(a) *The characteristic of* K *is a prime* p, K *is a finite dimensional vector space of dimension* s *over* $\mathbf{F}_p$, *and* $q = p^s$.
(b) *The multiplicative group* K^* *is cyclic of order* $q-1$.
(c) $x^{q-1} = 1$ *for every* $x \in \text{K}^*$; $x^q = x$ *for every* $x \in \text{K}$.

Proof. Since $\mathbf{Z}$ is infinite, K cannot be of characteristic zero. Thus K contains $\mathbf{F}_p$, p prime; in fact K is a vector space over $\mathbf{F}_p$, whose dimension s must be finite—otherwise K would be an infinite field. As a vector space, K is isomorphic to $(\mathbf{F}_p)^s$, so K contains p^s elements. We see that (b) follows from § 6, Theorem 1, and that (c) is an immediate consequence of (b).

Example. Let us interpret (b) for the case of $\mathbf{F}_p$, p prime. There exists an integer $x \in \mathbf{Z}$ such that $0 < x \leq p-1$ and such that every integer y which is not a multiple of p is congruent modulo p to a power of x. Such an x is called a *primitive root modulo* p. The problem of finding primitive roots modulo p is by no means trivial. For instance there are $\varphi(6) = 2$ roots primitive modulo 7; they are 3 and 5 (one sees that $1^2 \equiv 6^2 \equiv 1 \pmod 7$) and $2^3 \equiv 4^3 \equiv 1 \pmod 7$; 3 and 5 are the only other possibilities).

Remark. (a) and (c) imply that a finite field K with q elements is the set of roots of the polynomial $\text{X}^q - \text{X}$ (which has exactly q roots). One can show that two finite fields with q elements are isomorphic. We write $\mathbf{F}_q$ unambiguously for a field with q elements.

As an exercise we are going to digress in order to prove the following elegant theorem which concerns diophantine equations over a finite field.

Theorem 2 (Chevalley). *Let* K *be a finite field and let* $\text{F}(\text{X}_1, \ldots, \text{X}_n)$ *be a homogeneous polynomial of degree* d *over* K. *Suppose* $d < n$. *Then there exists a point* $(x_1, \ldots, x_n) \in \text{K}^n$ *distinct from the origin* $(0, \ldots, 0)$ *such that* $\text{F}(x_1, \ldots, x_n) = 0$.

Given a field K and an integer j, one says that K is a C_j*-field* if every homogeneous polynomial over K of degree d in n variables, such that $n > d^j$, has a non-trivial zero (i.e. besides the origin) in K^n. The C_0-fields are the algebraically closed fields. Chevalley's theorem asserts that finite fields are C_1 (one also calls C_1 fields "quasi-algebraically closed"). One can show that, if K is a C_j-field, the field $\text{K}(t)$ of rational functions in one variable over K and the field $\text{K}((t))$ of formal power series in one variable over K are C_{j+1}-fields ([5]). It had been, for a long time, an open question whether p-adic fields are C_2. Recently this has been shown not to be the case ([8]).

Proof of Theorem 2. Let us write q for card(K) and p for the characteristic of K (so $q = p^s$). Let $\text{V} \subset \text{K}^n$ be the set of zeros of F, i.e. the points $(x_1, \ldots, x_n) \in \text{K}^n$ such that $\text{F}(x) = 0$ (we use, hereafter, the symbol x to stand for a point $(x_1, \ldots, x_n) \in \text{K}^n$). According to Theorem 1, (c), $\text{F}(x)^{q-1} = 0$ for $x \in \text{V}$ and $\text{F}(x)^{q-1} = 1$ for $x \in \text{K}^n - \text{V}$. Thus the polynomial $\text{G}(x) = \text{F}(x)^{q-1}$ is the *characteristic function of* $\text{K}^n - \text{V}$ with values in

$\mathbf{F}_p$. The number modulo p of points of $K^n - V$ will thus be given by the sum $\sum_{x \in K^n} G(x)$. We are going to calculate this sum and show that it is zero. It will follow that card$(K^n - V)$ is a multiple of p; inasmuch as card $(K^n) = q^n = p^{ns}$ is also a multiple of p, card(V) will also have to be a multiple of p. Certainly V contains the origin, so, if $p \mid$ card(V), V necessarily contains other points, $p \geqslant 2$. Thus, to prove Theorem 2, it suffices to show that $\sum_{x \in K^n} G(x) = 0 \in \mathbf{F}_p$. Now, to calculate $\sum_{x \in K^n} G(x)$, we observe that the polynomial G is a linear combination of monomials $M_\alpha(X) = X_1^{\alpha_1} \ldots X_n^{\alpha_n}$. To determine $\sum_{x \in K^n} G(x)$ it suffices to calculate

$$\sum_{x \in K^n} M_\alpha(x) = \sum_{x \in K^n} x_1^{\alpha_1} \ldots x_n^{\alpha_n} = \Big(\sum_{x_1 \in K} x_1^{\alpha_1}\Big) \cdots \Big(\sum_{x_n \in K} x_n^{\alpha_n}\Big).$$

The problem reduces to that of calculating sums of the form $\sum_{z \in K} z^\beta$ $(\beta \in \mathbf{N})$.

(a) For $\beta = 0$, $z^\beta = 1$ for all $z \in K$. Consequently, $\sum_{z \in K} z^\beta = \sum_{z \in K} 1 = q = 0$.

(b) For $\beta > 0$, the term 0^β is 0, so the sum reduces to $\sum_{z \in K^*} z^\beta$. K^* is a cyclic group of order $q - 1$ (Theorem 1, (*b*)). Let ω generate K^*. Then $\sum_{z \in K^*} z^\beta = \sum_{j=0}^{q-2} \omega^{\beta j}$, which is the sum of a geometric progression. Thus:

(b′) If $\omega^\beta \neq 1$, i.e. if β is not a multiple of $q - 1$, then

$$\sum_{j=0}^{q-2} \omega^{\beta j} = \frac{\omega^{\beta(q-1)} - 1}{\omega^\beta - 1} = 0,$$

since $\omega^{q-1} = 1$.

(b″) If $\omega^\beta = 1$, i.e. if β is a multiple of $q - 1$, then

$$\sum_{j=0}^{q-2} \omega^{\beta j} = q - 1.$$

If follows from (a), (b′), and (b″) that $\sum_{x \in K^n} x_1^{\alpha_1} \ldots x_n^{\alpha_n}$ vanishes unless all the α_i's are non-zero and multiples of $q - 1$. In this case, the degree $\alpha_1 + \cdots + \alpha_n$ of the monomial is at least $(q - 1)n$. However, since $G = F^{q-1}$, G has degree $(q - 1)d$ and $(q - 1)d < (q - 1)n$ by assumption. Thus $\sum_{x \in K^n} M_\alpha(x) = 0$ for every monomial $M_\alpha(X)$ which appears in G with a non-zero coefficient. Therefore, $\sum_{x \in K^n} G(x) = 0$. We have seen that this relation implies the theorem.

Let us remark that it would have been sufficient, in place of the assumption that F was homogeneous, to have considered F with no constant term. Naturally, the strict inequality $d < n$ between the degree and the number of variables is essential. For example, the norm of $\mathbf{F}_{q^n}$ to $\mathbf{F}_q$ (cf. Chapter II, § 6) is a homogeneous polynomial of degree n in n variables over $\mathbf{F}_q$ with no non-trivial zero.

Example. A quadratic form in three variables over a finite field "represents zero" (i.e. has a non-trivial zero). Passing from K^3 to the projective plane $P_2(K)$, this means that a conic over K contains a point rational over K (i.e. whose coordinates may be chosen in K). The example of the conic $X^2 + Y^2 + Z^2 = 0$ over **R** (respectively $X^2 + Y^2 - 3Z^2 = 0$ over **Q**; in order to see that $X^2 + Y^2 - 3Z^2 = 0$ has no non-trivial solution in **Q**, it suffices to consider the case where x, y, z are relatively prime integers and then to reduce mod 4) shows that we are not dealing with a property common to all fields.

Chapter II

Elements integral over a ring; elements algebraic over a field

Among the complex numbers those which will concern us in this book are the so-called *algebraic numbers*, that is, those which satisfy an equation of the form

$$x^n + a_{n-1}x^{n-1} + \cdots + a_1 x + a_0 = 0,$$

where the coefficients are rational numbers. When the coefficients are integers ($a_i \in \mathbf{Z}$) the algebraic number x is called an *algebraic integer*. Thus $\sqrt{2}$, $\sqrt{3}$, i, $e^{2i\pi/5}$ are algebraic integers. It is not a priori clear that sums or products of algebraic numbers (respectively algebraic integers) are again algebraic numbers (respectively algebraic integers). Consider, for example, $x = \sqrt{2} + \sqrt{3}$. Squaring, one obtains $x^2 = 2 + 3 + 2\sqrt{6}$; adding and shifting across the equal sign gives $x^2 - 5 = 2\sqrt{6}$; again a squaring operation yields $(x^2 - 5)^2 = 24$, which shows that x is an algebraic integer. The reader will have to exert himself to show that $\sqrt[3]{5} + \sqrt[5]{7}$ is an algebraic integer and will be convinced that the sequence of steps which leads to a proof that this number is algebraic may not be easily generalized. In order to overcome this difficulty the algebraists of the last century, Dedekind in particular, had the idea of "linearizing" the problem, which means that they introduced the notion of module. We will begin with some results concerning modules. Considering modules over commutative rings will not require any extra effort and will be quite useful later. We will begin with the general case of integral elements over a ring and then specialize to algebraic elements over a field.

2.1. Elements integral over a ring

Theorem 1. Let R *be a ring,* A *a subring of* R, *and* x *an element of* R. *The following statements are equivalent:*

(a) *There exist* $a_0, \ldots, a_{n-1} \in \mathrm{A}$ *such that*

(1) $$x^n + a_{n-1}x^{n-1} + \cdots + a_1 x + a_0 = 0$$

(*i.e.* x *is a root of a monic polynomial with coefficients in* A).

(b) *The ring* $A[x]$ *is an A-module of finite type.*

(c) *There exists a subring* B *of* R *which contains* A *and* x *and which is an A-module of finite type.*

Proof. Call M the A-submodule of R generated by $1, x, \ldots, x^{n-1}$. By (a) $x^n \in M$. Multiplying (1) by x^j, we obtain $x^{n+j} = -a_{n-1}x^{n+j-1} - \cdots - a_0x^j$. In fact, induction on j implies that $x^{n+j} \in M$, for all $j \geq 0$. As $A[x]$ is the A-module generated by the x^k ($k \geq 0$), we see that $A[x] = M$. Thus (a) implies (b). That (b) implies (c) is clear. Let us show that (c) implies (a). Let $(y_1, \ldots, y_n)$ be a finite set of generators for B as a module over A, i.e. $B = Ay_1 + \cdots + Ay_n$. Since $x \in B$ and since B is a subring of R, it follows that $xy_i \in B$ for all $i = 1, \ldots, n$. Therefore,

$$xy_i = \sum_{j=1}^{n} a_{ij}y_j, \quad \text{for any } i = 1, \ldots, n, \quad a_{ij} \in A, \quad 1 \leq i, \quad j \leq n.$$

This means that

$$\sum_{j=1}^{n} (\delta_{ij}x - a_{ij})y_j = 0, i = 1, \ldots, n.$$

Consider this system of n homogeneous linear equations in $(y_1, \ldots, y_n)$. Write d for the determinant $\det(\delta_{ij}x - a_{ij})$. The calculation leading to Cramer's rule shows that $dy_i = 0$ for all i. This means that $db = 0$ for all $b \in B$; in particular, $d \cdot 1 = 0$, so $d = 0$. But d is clearly a monic polynomial in x, since the highest order term appears in the expansion of the product $\prod_{i=1}^{n} (x - a_{ii})$ of the entries on the principal diagonal. Thus (c) implies (a).

Definition 1. Let R *be a ring and let* A *be a subring of* R. *An element* x *of* R *is called integral over* A *if it satisfies the equivalent conditions* (a), (b), *and* (c) *of Theorem 1. Let* $P \in A[X]$ *be a monic polynomial such that* $P(x) = 0$ ((a) *implies that such a polynomial exists*). *The relation* $P(x) = 0$ *is called an equation of integral dependence of* x *over* A.

Example. The element $x = \sqrt{2}$ of $\mathbf{R}$ is integral over $\mathbf{Z}$. The relation $x^2 - 2 = 0$ is an equation of integral dependence.

Proposition 1. Let R *be a ring,* A *a subring of* R, *and let* $(x_i)_{1 \leq i \leq n}$ *be a finite set of elements of* R. *If, for all* i, x_i *is integral over* $A[x_1, \ldots, x_{i-1}]$ (*in particular if all the* x_i*'s are integral over* A), *then* $A[x_1, \ldots, x_n]$ *is an A-module of finite type.*

Proof. We argue by induction on n. For $n = 1$, we have a repeat of assertion (b) of Theorem 1. Assume that $B = A[x_1, \ldots, x_{n-1}]$ is an A-module of finite type. Then $B = \sum_{j=1}^{p} Ab_j$. The case $n = 1$ implies that $A[x_1, \ldots, x_n] = B[x_n]$ is a B-module of finite type. Write $B[x_n] = \sum_{k=1}^{q} Bc_k$. Then

$$A[x_1, \ldots, x_n] = \sum_{k=1}^{q} Bc_k = \sum_{k=1}^{q} \left(\sum_{j=1}^{p} Ab_j\right)c_k = \sum_{j,k} Ab_jc_k.$$

Thus $(b_jc_k)_{\substack{1 \leq j \leq p \\ 1 \leq k \leq q}}$ is a finite set of generators for $A[x_1, \ldots, x_n]$ as a module over A.

Corollary 1. Let R *be a ring,* A *a subring of* R, x *and* y *elements of* R *which are integral over* A. *Then* $x + y$, $x - y$, *and* xy *are integral over* A.

Proof. Clearly $x + y$, $x - y$, and $xy \in A[x, y]$. According to Proposition 1 $A[x, y]$ is an A-module of finite type. According to part (c) of Theorem 1, $x + y$, $x - y$, and xy are integral over A.

Corollary 2. Let R *be a ring and let* A *be a subring of* R. *The set* A′ *of elements of* R *which are integral over* A *is a subring of* R *which contains* A.

Proof. Corollary 1 implies that A′ is a subring of R. We have $A \subset A'$, since, if $a \in A$, a is a root of the monic polynomial $P(X) = X - a$, which has coefficients in A.

Definition 2. Let R *be a ring,* A *a subring of* R. *The ring* A′ *of elements of* R *which are integral over* A *is called the integral closure of* A *in* R. *Let* A *be an integral domain and let* K *be its field of fractions. The integral closure of* A *in* K *is called the integral closure of* A. *Let* B *be a ring and* A *a subring of* B. *We say that* B *is integral over* A *if every element of* B *is integral over* A (*i.e. if the integral closure of* A *in* B *is* B *itself*).

Proposition 2 (Transitivity). *Let* C *be a ring,* B *a subring of* C, *and* A *a subring of* B. *If* B *is integral over* A *and if* C *is integral over* B, *then* C *is integral over* A.

Proof. Let $x \in C$. Then x is integral over B, so there is an equation of integral dependence $x^n + b_{n-1}x^{n-1} + \cdots + b_0 = 0$ with $b_i \in B, i = 0, \ldots, n-1$. Put $B' = A[b_0, \ldots, b_{n-1}]$. Then x is integral over B′. As B is integral over A, the b_i are integral over A. Therefore, Proposition 1 implies that $B'[x] = A[b_0, \ldots, b_{n-1}, x]$ is an A-module of finite type. By part (c) of Theorem 1, x is integral over A. Thus C is integral over A.

Proposition 3. Let B *be an integral domain and* A *a subring of* B *such that* B *is integral over* A. *In order that* B *be a field it is necessary and sufficient that* A *be a field.*

Proof. Suppose that A is a field and let $b \in B$, $b \neq 0$. Then $A[b]$ is a *finite* dimensional vector space over A (part (*b*) of Theorem 1). On the other hand $y \mapsto by$ is an A-linear transformation of $A[b]$. It is injective since $A[b]$ is an integral domain and since $b \neq 0$. Therefore, it is surjective. There exists $b' \in A[b]$ such that $bb' = 1$. This means that, for any $b \in B - (0)$, b is invertible in B, so B is a field.[1]

Conversely, suppose that B is a field. Let $a \in A - (0)$. Then a has an inverse $a^{-1} \in B$ which satisfies an equation of integral dependence

$$a^{-n} + a_{n-1}a^{-n+1} + \cdots + a_1a^{-1} + a_0 = 0, \quad a_i \in A.$$

Multiplying by a^{n-1}, we obtain

$$a^{-1} = -(a_{n-1} + \cdots + a_1a^{n-2} + a_0a^{n-1}),$$

which shows that $a^{-1} \in A$. Thus A is a field.

---

1. The same reasoning, involving the mapping $y \mapsto by$, allows one to conclude that any *finite* integral domain is a field.

2.2. Integrally closed rings

Definition. A ring A *is called integrally closed if it is an integral domain and if it is its own integral closure.*

In other words, every element x of the field of fractions K of A which is integral over A belongs to A.

Example 1. Let A be an integral domain and let K be its field of fractions. Then the *integral closure* A′ of A (i.e. the integral closure of A in K) is integrally closed. This follows from the fact that the integral closure of A′ is integral over A′, therefore over A(§ 1, Proposition 2). It therefore equals A′.

Example 2. Every principal ideal ring is integrally closed.

Proof. By definition a principal ideal ring is an integral domain. Let x be an element of the field of fractions of A which is integral over A. Let

$$x^n + a_{n-1}x^{n-1} + \cdots + a_1x + a_0 = 0 \quad (a_i \in \mathrm{A}) \tag{1}$$

be an integral dependence equation for x over A. Write $x = a/b$ with a and b *relatively prime* elements of A. Substitute in (1) and multiply through by b^n to obtain

$$a^n + b(a_{n-1}a^{n-1} + \cdots + a_1ab^{n-2} + a_0b^{n-1}) = 0.$$

Thus b divides a^n. As b is relatively prime to a, repeated application of Euclid's lemma shows that b divides a. Therefore, b is a unit in A. Thus, $x = a/b \in \mathrm{A}$ and A is integrally closed.

One may observe that only the multiplicative properties of principal ideal rings have been used (relative primeness, Euclid's lemma). The same argument thus shows that every factorial ring is integrally closed.

2.3. Elements algebraic over a field. Algebraic extensions

Definition. Let R *be a ring and* K *a subfield of* R. *An element* $x \in \mathrm{R}$ *is called algebraic over* K *if there exist elements* $a_0, \ldots, a_n \in \mathrm{K}$, *not all of which are zero, such that* $a_nx^n + \cdots + a_1x + a_0 = 0$.

Equivalently, the monomials $(x^j)_{j\in\mathrm{N}}$ are linearly dependent over K. An element of R which is not algebraic over K is called *transcendental* over K; i.e. x is transcendental over K if and only if the monomials $(x^j)_{j\in\mathrm{N}}$ are linearly independent over K.

In the relation of Definition 1, we may assume that $a_n \neq 0$. In this case $a_n^{-1} \in \mathrm{K}$; multiplying through by a_n^{-1} we obtain an equation of integral dependence. Therefore:

$$\textit{Over a field, algebraic} = \textit{integral.} \tag{1}$$

We may thus apply the theory of integral elements. For example, for $\mathrm{K} \subset \mathrm{R}$ and $x \in \mathrm{R}$, Theorem 1, (b) of § 1 asserts:

$$x \textit{ algebraic over } \mathrm{K} \Leftrightarrow [\mathrm{K}[x] : \mathrm{K}] \textit{ finite.} \tag{2}$$

We say that a ring R containing a field K is *algebraic* over K if every element of R is algebraic over K. If R is a field, then R is called an *algebraic extension* of K.

Given a field L and a subfield K of L, we call the dimension $[L:K]$ the *degree* of L over K. In this context Theorem 1, (c) of § 1 has the following interpretation:

(3) *If the degree of* L *over* K *is finite,* L *is an algebraic extension of* K.

Any extension field of finite degree over $\mathbf{Q}$ is called an *algebraic number field* (or, simply a *number field*).

Proposition 1. Let K *be a field,* L *an algebraic extension of* K, *and* M *an algebraic extension of* L. *Then* M *is an algebraic extension of* L. *Furthermore,* $[M:K] = [M:L][L:K]$ ("*multiplicativity of degrees*").

Proof. The first assertion is a special case of Proposition 2 of § 1. Moreover, if $(x_i)_{i\in I}$ is a base of L over K and $(y_j)_{j\in J}$ is a base for M over L, then $(x_iy_j)_{(i,j)\in I\times J}$ is a base for M over K. As in Proposition 1 of § 1 we see that $(x_iy_j)_{(i,j)\in I\times J}$ generates M over K. A relation $\sum a_{ij}x_iy_j = 0$ with $a_{ij}\in K$ entails $\sum (\sum a_{ij}x_i)y_j = 0$, whence $\sum a_{ij}x_i = 0$ for all j (since $\sum a_{ij}x_i \in L$), and consequently $a_{ij} = 0$ for all $(i, j) \in I \times J$. This proves that $[M:K] = [M:L]\,[L:K]$.

Proposition 2. Let R *be a ring and* K *a subfield of* R. *Then:*

(a) *The set* K′ *of elements of* R *algebraic over* K *is a subring of* R *containing* K.
(b) *If* R *is an integral domain,* K′ *is a subfield of* R.

Proof. (a) is a special case of Corollary 2 of Proposition 1, § 1, and (b) follows from Proposition 3 of § 1. Q.E.D.

Now we study the elements algebraic over a field in greater detail. Let R be a ring, K a subfield of R, and let x be an element of R. Write K[X] for the ring of polynomials in one variable over K. There exists a unique homomorphism $\varphi: K[X] \to R$ such that $\varphi(X) = x$ and such that $\varphi(a) = a$ for all $a \in K$. The image of φ is $K[x]$. The definition of algebraic element may be reformulated as follows:

(4) *An element x is algebraic over* $K \Leftrightarrow \mathrm{Ker}(\varphi) \neq (0)$.

Proof. If x is transcendental over K, then obviously $\mathrm{Ker}(\varphi) = (0)$. In any case the ideal $\mathrm{Ker}(\varphi)$ is a principal ideal $(F(X))$ (since K[X] is a principal ideal ring). In the case that x is algebraic over K, it is generated by a non-zero polynomial $F(X)$.

Remark. We may assume that $F(X)$ is monic, since K is a field. $F(X)$ is then uniquely determined by K and x; we call it the *minimal polynomial* of x over K. Its properties are as follows:

(5) *Let* $F(X)$ *be the minimal polynomial of x over* K. *Let* $G(X) \in K[X]$. $G(x) = 0$ *if and only if* $F(X)$ *divides* $G(X)$ *in* $K[X]$.

Passing to the quotient ring, we obtain a *canonical isomorphism:*

(6) $$K[X]/(F(X)) \xrightarrow{\sim} K[x].$$

With the same notations, suppose now that x is algebraic over K and let $F(X)$ be its minimal polynomial. Applying (5) and Proposition 3 of § 1, we obtain the equivalence of the following statements:

(7) $K[x]$ *is a field.* $\Leftrightarrow$ $K[x]$ *is an integral domain.* $\Leftrightarrow$ $F(X)$ *is irreducible.*

On the other hand, if K is a field and $F(X) \in K[X]$ is irreducible, then $K[X]/(F(X))$ is a field containing K and, writing x for the projection of $X \in K[X]$ into this field, we have $F(x) = 0$. Thus $X - x$ divides $F(X)$ in the field $K[x]$. More generally:

Proposition 3. Let K *be a field and let* $P(X) \in K[X]$ *be a nonconstant polynomial. There exists an algebraic extension of finite degree* K' *of* K *such that* $P(X)$ *factors in* $K'[X]$ *into a product of polynomials of degree one* (*linear polynomials*).

Proof. We argue by induction on the degree $d°$ of $P(X)$. There is nothing to prove in the case $d° = 1$. Let $F(X)$ be an irreducible factor of $P(X)$. We have just seen that there exists an extension K'' of finite degree over K (e.g. $K[X]/(F(X))$) containing an element x such that $X - x$ divides $F(X)$ in $K''[X]$. Thus $P(X) = (X - x)P_1(X)$ with $P_1(X) \in K''[X]$. According to the induction hypothesis $P_1(X)$ factors into a product of linear polynomials in an extension K' of finite degree over K''. By Proposition 1, K' is of finite degree over K, and $P(X)$ is a product of linear polynomials in $K'[X]$.

Remark. Algebraically closed fields. A field K is called *algebraically closed* if every non-constant polynomial $P(X) \in K[X]$ may be expressed as a product of linear factors, all lying in $K[X]$. For this, as follows by induction on the degree of the polynomial, it suffices that every non-constant polynomial have a root in K. Extending Proposition 3 by means of Zorn's lemma (cf. [1], Chapter V, and [9], Chapter II), one may show that any field is imbeddable in an algebraically closed field.

One may prove, by means of the techniques of mathematical analysis and in many different ways,[1] that the field **C** of complex numbers is algebraically closed. We shall not need more than this.

2.4. Conjugate elements, conjugate fields

Given two fields L and L' both containing a field K we call any isomorphism $\varphi: L \to L'$ such that $\varphi(a) = a$ for all $a \in K$ a K-*isomorphism* of L on L'. In this case we say that L and L' are K-*isomorphic*, or (if they are algebraic over K) we say that they are *conjugate over* K. Given two extensions L and L' of K, we say that two elements $x \in L$ and $x' \in L'$ are *conjugate* over K if there exists a K-isomorphism $\varphi: K(x) \to K(x')$ such that $\varphi(x) = x'$. Such a φ is, of course, unique. The existence of a φ means that either x and x' are both transcendental over K or both are algebraic over K with the same minimal polynomial (cf. (5), § 3).

Example. Let $F(X)$ be an irreducible polynomial of degree n over K and let $x_1, \ldots, x_n$

1. For a proof which depends upon the properties of continuous functions defined on compact topological spaces, see [4]. For a proof which makes use of the properties of holomorphic functions of a complex variable, see [3]. We will give in an appendix to this chapter a more algebraic proof, which will involve no analysis beyond the most elementary properties of the real numbers.

be its roots in an extension K′ of K (§ 3, Proposition 3). Then the x_i's are pairwise conjugate over K (§ 3, (6)), and the fields $K[x_i]$ are also pairwise conjugate.

Lemma. Let K *be a field of characteristic zero or a finite field, let* $F(X) \in K[X]$ *be a monic irreducible polynomial, and let* $F(X) = \prod_{i=1}^{n} (X - x_i)$ *be its decomposition into a product of linear factors in an extension* K′ *of* K (§ 3, Proposition 3). *Then the n roots* $x_1, \ldots, x_n$ *of* F(X) *are distinct.*

Proof. If not, F(X) has a root in common with its derivative F′(X). Therefore, F(X) divides F′(X) ((4), § 3). Since $d°F' < d°F$, this means that F′(X) is the polynomial zero. However, $F(X) = X^n + a_{n-1}X^{n-1} + \cdots + a_0$ $(a_i \in K)$ and

$$F'(X) = nX^{n-1} + (n-1)a_{n-1}X^{n-2} + \cdots + a_1.$$

Thus $n \cdot 1 = 0, j \cdot a_j = 0, j = 1, \ldots, n-1$, which is impossible in characteristic zero. In characteristic $p \neq 0$, these relations imply that p divides n and that $a_j = 0$ for all j not divisible by p (recall that p is a prime number). Thus F(X) is of the form

$$F(X) = X^{qp} + b_{q-1}X^{(q-1)p} + \cdots + b_1X^p + b_0 \quad (b_i \in K).$$

If each of the b_i's is a pth power, i.e. $b_i = c_i^p$ with $c_i \in K$, then

$$F(X) = (X^q + c_{q-1}X^{q-1} + \cdots + c_0)^p \quad \text{(Chapter I, § 7, Proposition 1)},$$

and F(X) is not irreducible. But, if K is a finite field with $p \neq 0$ its characteristic, the mapping $x \mapsto x^p$ of K into K is injective (since $x^p = y^p$ implies $x^p - y^p = 0$, so $(x-y)^p = 0$ which implies $x - y = 0$); it is thus surjective, since K is finite. Therefore, F(X) is not irreducible and we have a contradiction.

The fields K of characteristic $p \neq 0$ for which $x \mapsto x^p$ is surjective (i.e. for which every element of K is a pth power) are called *perfect* fields. We have just shown that finite fields are perfect. By convention fields of characteristic zero are also considered perfect. The preceding lemma is true for K any perfect field (and our proof works in this more general case). The field $\mathbf{F}_p(T)$ of rational functions in one variable over $\mathbf{F}_p$ is not perfect, inasmuch as the variable T is not a pth power in $\mathbf{F}_p(T)$.

Theorem 1. Let K *be a field of characteristic zero or a finite field, let* K′ *be an extension of finite degree n of* K, *and let* C *be an algebraically closed field containing* K. *Then there exist n distinct* K*-isomorphisms of* K′ *into* C.

Proof. Our assertion is true for any extension field K′ of K which is of the form $K[x]$ with $x \in K'$. In fact, the minimal polynomial F(X) of x over K is then of degree n. It has n roots $x_1, \ldots, x_n$ in C, all of which are distinct according to the lemma. For any $i = 1, \ldots, n$ we have then a K-isomorphism $\sigma_i : K' \to C$ such that $\sigma_i(x) = x_i$. We continue by induction on the degree n of K′. Let $x \in K'$, consider the fields $K \subset K[x] \subset K'$ and put $q = [K[x] : K]$. We may assume $q > 1$. We have seen that there are q distinct K-isomorphisms $\sigma_1, \ldots, \sigma_q$ of $K[x]$ into C. As $K[\sigma_i(x)]$ and $K[x]$ are isomorphic, it is possible to construct an extension K_i' of $K[\sigma_i(x)]$ and an isomorphism $\tau_i : K' \to K_i'$ which extends σ_i. Clearly $K[\sigma_i(x)]$ is a field of characteristic zero or a finite field. Since

$[K'_i : K[\sigma_i(x)]] = [K' : K[x]] = n/q < n$, the induction hypothesis implies that there are n/q distinct $K[\sigma_i(x)]$-isomorphisms θ_{ij} of K'_i into C. Therefore, the n composed mappings $\theta_{ij} \circ \tau_i$ provide $q \cdot n/q = n$ K-isomorphisms of K′ into C. They are distinct, since, for $i \neq i'$, $\theta_{ij} \circ \tau_i$ and $\theta_{i'j'} \circ \tau_{i'}$ differ on $K[x]$ while, for $i = i'$ but $j \neq j'$, θ_{ij} and $\theta_{ij'}$ differ on K'_i. Q.E.D.

Theorem 1 extends to the case of a perfect field K. One shows that any algebraic extension of a perfect field (in particular $K[\sigma_i(x)]$) is a perfect field. The rest of the proof remains unchanged.

Corollary ("theorem of the primitive element"). *Let* K *be a finite field or a field of characteristic zero. Let* K′ *be an extension of* K *of finite degree n. Then there exists an element x of* K′ *(called a "primitive element") such that* $K' = K[x]$.

Proof. If K is finite, K′ is finite and its multiplicative group K'^* is comprised of the powers of a single element x (Chapter I, § 7, Theorem 1, (b)). Thus $K' = K[x]$. Suppose that K is of characteristic zero and thus an infinite field. According to Theorem 1 there are n K-isomorphisms σ_i of K′ into an algebraically closed field C containing K. For $i \neq j$ the equation $\sigma_i(y) = \sigma_j(y)$ $(y \in K')$ defines a subset V_{ij} of K′, which is clearly a K-subspace of the vector space K′ and which is distinct from K′ when $\sigma_i \neq \sigma_j$. Since K is infinite, linear algebra shows that the union of the V_{ij} is not all of K′. Take x outside this union. The $\sigma_i(x)$ are then pairwise distinct, so that the minimal polynomial F(X) of x over K has at least n distinct roots (the $\sigma_i(x)$) in C. Therefore, $d°F \geq n$, i.e. $[K[x] : K] \geq n$. Since $K[x] \subset K'$ and since $[K' : K] = n$, we conclude that $K' = K[x]$. Q.E.D.

2.5. Integers in quadratic fields

We pause a moment from the development of the general theory to give an example.

Definition. Any extension field of degree 2 over the field **Q** *of rational numbers is called a quadratic field.*

If K is a quadratic field, any element $x \in K - \mathbf{Q}$ is of degree 2 over **Q**, thus is a primitive element of K (i.e. $K = \mathbf{Q}[x]$ and $(1, x)$ is a base of K over **Q**). Let $F(X) = X^2 + bX + c$ $(b, c \in \mathbf{Q})$ be the minimal polynomial of such an element $x \in K$. Solving the quadratic equation $x^2 + bx + c = 0$ gives $2x = -b \pm \sqrt{b^2 - 4c}$. Thus $K = \mathbf{Q}(\sqrt{b^2 - 4c})$.[1] Now $b^2 - 4c$ is a rational number $u/v = uv/v^2$ with $u, v \in \mathbf{Z}$. One sees that $K = \mathbf{Q}(\sqrt{uv})$ with $u, v \in \mathbf{Z}$. In fact, one sees that it is possible to write $K = \mathbf{Q}(\sqrt{d})$ where d is a square-free integer (d is plus or minus a product of distinct primes). Thus:

Proposition 1. Every quadratic field is of the form $\mathbf{Q}(\sqrt{d})$, *where d is a square-free integer.*

The element $\sqrt{d}$ is a root of the irreducible polynomial $X^2 - d$. This element $\sqrt{d}$ has a *conjugate* in K, and the conjugate has to be $-\sqrt{d}$. There exists an automorphism

1. By $\sqrt{b^2 - 4c}$ we mean one of the two elements of K whose square is $b^2 - 4c$.

σ of K which sends $\sqrt{d}$ to $-\sqrt{d}$ (§ 4). Any element of K is of the form $a + b\sqrt{d}$ with $a, b \in \mathbf{Q}$. We have

$$\sigma(a + b\sqrt{d}) = a - b\sqrt{d}. \tag{1}$$

We consider the ring A of *integers* of K, i.e. the set of $x \in$ K which are integral over $\mathbf{Z}$ (§ 1, Corollary 2 of Proposition 1). If $x \in$ A, $\sigma(x)$ is a root of the same equation of integral dependence as x, so $\sigma(x) \in$ A. We have then that $x + \sigma(x) \in$ A and $x \cdot \sigma(x) \in$ A. But, if $x = a + b\sqrt{d}$ with $a, b \in \mathbf{Q}$, then, according to (1),

$$x + \sigma(x) = 2a \in \mathbf{Q} \quad \text{and} \quad x \cdot \sigma(x) = a^2 - db^2 \in \mathbf{Q}. \tag{2}$$

Since $\mathbf{Z}$ is a principal ideal ring and hence integrally closed (§ 2, Example 2), we see that

$$2a \in \mathbf{Z}; \qquad a^2 - db^2 \in \mathbf{Z}. \tag{3}$$

The conditions (3) are necessary in order that $x = a + b\sqrt{d}$ be integral over $\mathbf{Z}$. They are also sufficient, since x is a root of $X^2 - 2aX + a^2 - db^2 = 0$. According to (3), $(2a)^2 - d(2b)^2 \in \mathbf{Z}$. Since $2a \in \mathbf{Z}$, we have $d(2b)^2 \in \mathbf{Z}$ too. On the other hand, d is square-free, so, if $2b$ were not an integer, its denominator would have to include a prime factor p. This prime factor would have to appear as p^2 in the denominator of $(2b)^2$. Multiplication by d would not send $(2b)^2$ into $\mathbf{Z}$. We may conclude that $2b \in \mathbf{Z}$.

In brief, we may take $a = u/2$, $b = v/2$ with $u, v \in \mathbf{Z}$. Condition (3) becomes:

$$u^2 - dv^2 \in 4\mathbf{Z}. \tag{4}$$

If v is even, (4) shows that u is even too. In this case, $a, b \in \mathbf{Z}$. If v is odd, then $v^2 \equiv 1$ mod 4. The possibilities mod 4 for u^2 are 0 and 1 (the only squares mod 4). Since d is square-free, it is not a multiple of 4. Necessarily $u^2 \equiv 1$ mod 4 and $d \equiv 1$ mod 4. We have proved the following:

Theorem 1. Let $K = \mathbf{Q}(\sqrt{d})$ *be a quadratic field with* $d \in \mathbf{Z}$ *square-free (therefore* $\not\equiv 0$ mod 4).

(a) *If* $d \equiv 2$ *or* $d \equiv 3$ mod 4, *the ring* A *of integers of* K *consists of all elements of the form* $a + b\sqrt{d}$ *with* $a, b \in \mathbf{Z}$.

(b) *If* $d \equiv 1$ mod 4, A *consists of all elements of the form* $\frac{1}{2}(u + v\sqrt{d})$ *with* u *and* $v \in \mathbf{Z}$ *of the same parity.*

In the case that $d \equiv 2$ or 3 mod 4, $(1, \sqrt{d})$ is a base for A as a $\mathbf{Z}$-module. If $d \equiv 1$ mod 4, $(1, \frac{1}{2}(1 + \sqrt{d}))$ is a $\mathbf{Z}$-module base for A. Indeed, by b), 1 and $\frac{1}{2}(1 + \sqrt{d})$ belong to A. Conversely, to show that $\frac{1}{2}(u + v\sqrt{d})$ (with $u, v \in \mathbf{Z}$ of the same parity) is expressible as a $\mathbf{Z}$-linear combination of 1 and $\frac{1}{2}(1 + \sqrt{d})$, one may, by subtracting $\frac{1}{2}(1 + \sqrt{d})$, reduce the problem to the case where u and v are even. In this case

$$\tfrac{1}{2}(u + v\sqrt{d}) = \left(\frac{u}{2} - \frac{v}{2}\right) \cdot 1 + v \cdot \tfrac{1}{2}(1 + \sqrt{d}).$$

We conclude with some *terminology*. If $d > 0$, $\mathbf{Q}(\sqrt{d})$ is called a *real quadratic field* (there exists a subfield of $\mathbf{R}$ conjugate to $\mathbf{Q}(\sqrt{d})$ over $\mathbf{Q}$). If $d < 0$, then $\mathbf{Q}(\sqrt{d})$ is called an *imaginary quadratic field.*

2.6. Norms and traces

(a) Review of linear algebra

Let A be a ring, E a *free* A-module of finite rank and let u be an endomorphism of E. In linear algebra one defines the *trace*, the *determinant*, and the *characteristic polynomial* of u. If a base (e_i) of E has been chosen and if (a_{ij}) is the matrix for u with respect to this base, then the trace, determinant, and characteristic polynomial of a are, respectively,

$$(1)\quad \mathrm{Tr}(u) = \sum_{i=1}^{n} a_{ii}, \quad \det(u) = \det(a_{ij}), \quad \text{and} \quad \det(\mathrm{X}\cdot \mathrm{I_E} - u) = \det(\mathrm{X}\,\delta_{ij} - a_{ij}).$$

NB. These quantities are independent of the choice of base.

The formulas (1) imply:

$$(2)\qquad \mathrm{Tr}(u + u') = \mathrm{Tr}(u) + \mathrm{Tr}(u'),$$
$$\det(uu') = \det(u)\det(u'),$$
and
$$\det(\mathrm{X}\cdot \mathrm{I_E} - u) = \mathrm{X}^n - (\mathrm{Tr}(u))\mathrm{X}^{n-1} + \cdots + (-1)^n \det(u).$$

(b) Norms and traces in an extension

Let B be a ring and let A be a subring of B such that B is a free A-module of finite rank n (for example, A can be a field and B a finite extension of degree n of A). For $x \in \mathrm{B}$, multiplication m_x by x (i.e. $y \mapsto xy$) is an endomorphism of the A-module B.

Definition 1. We call trace (respectively, norm, characteristic polynomial) of $x \in \mathrm{B}$, relative to B *and* A, *the trace (respectively, determinant, characteristic polynomial) of the endomorphism m_x of multiplication by x.*

The trace (respectively, norm) of x is denoted $\mathrm{Tr_{B/A}}(x)$ (respectively, $\mathrm{N_{B/A}}(x)$), or $\mathrm{Tr}(x)$ (respectively, $\mathrm{N}(x)$) when no confusion is possible. They are elements of A. The characteristic polynomial is a monic polynomial with coefficients in A.

For $x, x' \in \mathrm{B}$ and $a \in \mathrm{A}$ we have $m_x + m_{x'} = m_{x+x'}$ and $m_x \circ m_{x'} = m_{xx'}$ and $m_{ax} = am_x$. Furthermore, the matrix of m_a with respect to any base for B over A is the diagonal matrix all of whose diagonal entries are a. From formulas (1) and (2) we obtain:

$$(3)\qquad \mathrm{Tr}(x + x') = \mathrm{Tr}(x) + \mathrm{Tr}(x'), \quad \mathrm{Tr}(ax) = a\,\mathrm{Tr}(x), \quad \mathrm{Tr}(a) = n\cdot a$$
$$\mathrm{N}(xx') = \mathrm{N}(x)\,\mathrm{N}(x'), \quad \mathrm{N}(a) = a^n, \quad \text{and} \quad \mathrm{N}(ax) = a^n\mathrm{N}(x).$$

Proposition 1. Let K *be a field of characteristic* 0 *or a finite field, let* L *be an algebraic extension of degree n of* K, *let x be an element of* L, *and let $x_1, \ldots, x_n$ be the roots of the minimal polynomial of x over* K *(in a suitable extension of* K; *cf.* § 3, *Proposition* 3), *each one repeated* $[\mathrm{L} : \mathrm{K}[x]]$ *times. Then* $\mathrm{Tr_{L/K}}(x) = x_1 + \cdots + x_n$, $\mathrm{N_{L/K}}(x) = x_1 \cdots x_n$. *The characteristic polynomial of x, relative to* L *and* K *is* $(\mathrm{X} - x_1) \ldots (\mathrm{X} - x_n)$.

Thus the characteristic polynomial is the $[\mathrm{L} : \mathrm{K}[x]]$th power of the minimal polynomial of x over K.

Proof. Let us first treat the case where x is a *primitive element* of L over K (cf. § 4, corollary of Theorem 1). Let $F(X)$ be the minimal polynomial of x over K. Then L is K-isomorphic to $K[X]/(F(X))$ (§ 3, formula (5)), and $(1, x, \ldots, x^{n-1})$ is a base for L over K. Let us put $F(X) = X^n + a_{n-1}X^{n-1} + \cdots + a_0$. The matrix of the endomorphism m_x with respect to this base is

$$M = \begin{vmatrix} 0 & 0 & \ldots & 0 & -a_0 \\ 1 & 0 & \ldots & 0 & -a_1 \\ 0 & 1 & \ldots & 0 & \vdots \\ \vdots & 0 & & \vdots & \vdots \\ \vdots & \vdots & & \vdots & \vdots \\ 0 & 0 & \ldots & 1 & -a_{n-1} \end{vmatrix}.$$

The determinant of $X \cdot I_L - m_x$ is therefore the determinant of the matrix

$$X \cdot I_n - M = \begin{vmatrix} X & 0 & \ldots & 0 & a_0 \\ -1 & X & & 0 & a_1 \\ 0 & -1 & & 0 & \vdots \\ \vdots & 0 & & \vdots & \vdots \\ \vdots & \vdots & & X & a_{n-2} \\ 0 & 0 & & -1 & X + a_{n-1} \end{vmatrix}.$$

Expanding this determinant as a polynomial in X, we obtain the characteristic polynomial of x. It is clearly equal to $F(X)$, the minimal polynomial of x. By (2), $\mathrm{Tr}(x) = -a_{n-1}$ and $N(x) = (-1)^n a_0$. Since x is primitive, $F(X) = (X - x_1) \ldots (X - x_n)$; equating coefficients we see that $\mathrm{Tr}(x) = x_1 + \cdots + x_n$ and $N(x) = x_1 \ldots x_n$.

Consider now the general case. Put $r = [L : K[x]]$. It suffices to show that the characteristic polynomial $P(X)$ of x, with respect to L and K, is equal to the rth power of the minimal polynomial of x over K. Let $(y_i)_{i=1,\ldots,q}$ be a base for $K[x]$ over K and let $(z_j)_{j=1,\ldots,r}$ be a base for L over $K[x]$; then $(y_i z_j)$ is a base for L over K and $n = qr$ (§ 3, Proposition 1). Let $M = (a_{ih})$ be the matrix for multiplication by x in $K[x]$ with respect to the base (y_i): thus $xy_i = \sum_h a_{ih} y_h$. We have then

$$x(y_i z_j) = \left(\sum_h a_{ih} y_h\right) z_j = \sum_h a_{ih}(y_h z_j).$$

Ordering lexicographically the base $(y_i z_j)$ of L over K, we see that the matrix M_1 for multiplication by x in L with respect to this base takes the form

$$M_1 = \begin{vmatrix} M & 0 & \ldots & 0 \\ 0 & M & \ldots & 0 \\ \vdots & & \vdots\vdots\vdots & \vdots \\ 0 & 0 & \ldots & M \end{vmatrix},$$

i.e. M occurring r-times as diagonal blocks in M_1. The matrix $X \cdot I_n - M_1$ thus consists of r diagonal blocks, each of the form $X \cdot I_q - M$. Consequently, $\det(X \cdot I_n - M_1) = (\det(X \cdot I_q - M))^r$. The left-hand side of the preceding equation is $P(X)$; $\det(X \cdot I_q - M)$ is the minimal polynomial of x over K, according to the first part of the proof. Q.E.D.

In conclusion we present a result regarding traces and norms of integral elements.

Proposition 2. Let A *be an integral domain,* K *its field of fractions,* L *an extension of finite degree of* K, *and x an element of* L *integral over* A. *Assume* K *has characteristic zero. Then the coefficients of the characteristic polynomial* $P(X)$ *of x relative to* L *and* K, *in particular* $\mathrm{Tr}_{L/K}(x)$ *and* $N_{L/K}(x)$, *are integral over* A.

Proof. We make use of Proposition 1. We have $P(X) = (X - x_1) \ldots (X - x_n)$; the coefficients of $P(X)$ are thus, up to a sign, sums of products of the x_i's. It suffices to show that the x_i's are integral over A (§ 1, Corollary 1 of Proposition 1). But each x_i is a conjugate of x over K (§ 4), and there is a K-isomorphism $\sigma_i : K[x] \to K[x_i]$ such that $\sigma_i(x) = x_i$. Applying σ_i to an equation of integral dependence of x over A, we obtain an equation of integral dependence for x_i over A.

Corollary. Suppose, further, that A *is integrally closed. Then the coefficients of the characteristic polynomial of x, in particular* $\mathrm{Tr}_{L/K}(x)$ *and* $N_{L/K}(x)$, *are elements of* A.

Proof. By definition these coefficients are elements of K. By Proposition 2 they are integral over A. Q.E.D.

We remark that the quantities $x + \sigma(x)$ and $x \cdot \sigma(x)$ employed in the discussion of quadratic fields (§ 5) are the trace and the norm of x. We proved a special case of the above corollary (§ 5, (3)) in the course of our discussion of quadratic fields.

2.7. The discriminant

Definition 1. Let B *be a ring and let* A *be a subring of* B *such that* B *is a free* A*-module of finite rank n. For $(x_1, \ldots, x_n) \in B^n$ we call the discriminant of the set $(x_1, \ldots, x_n)$ the element of* A *defined by the relation*

$$D(x_1, \ldots, x_n) = \det(\mathrm{Tr}_{B/A}(x_i x_j)). \tag{1}$$

Proposition 1. If $(y_1, \ldots, y_n) \in B^n$ is another set of elements of B *such that $y_i = \sum_{j=1}^{n} a_{ij} x_j$ with $a_{ij} \in A$, then*

$$D(y_1, \ldots, y_n) = (\det(a_{ij}))^2 D(x_1, \ldots, x_n). \tag{2}$$

Proof.

$$\mathrm{Tr}(y_p y_q) = \mathrm{Tr}(\sum_{i,j} a_{pi} a_{qj} x_{ij}) = \sum_{i,j} a_{pi} a_{qj} \mathrm{Tr}(x_i y_j).$$

This gives the matrix equation: $(\mathrm{Tr}(y_p y_q)) = (a_{pi})(\mathrm{Tr}(x_i x_j)) \cdot {}^t(a_{qj})$ (where ${}^t M$ denotes the transpose of the matrix M). To complete the proof it suffices to take determinants. Q.E.D.

Proposition 1 implies that the discriminants of bases for B over A are *associates* in A; i.e. the matrix (a_{ij}) which expresses one base in terms of another has an inverse with entries in A. Therefore both $\det(a_{ij})$ and $\det(a_{ij})^{-1}$ are units in A. We may thus formulate the following definition:

Definition 2. Under the hypotheses of Definition 1 we call the principal ideal of A *generated by the discriminant of any base of* B *over* A *the discriminant of* B *over* A. *We denote it* $\mathscr{D}_{B/A}$.

Proposition 2. Suppose that $\mathscr{D}_{B/A}$ *contains an element which is not a zero-divisor. Then, in order that a set* $(x_1, \ldots, x_n) \subset B^n$ *be a base for* B *over* A, *it is necessary and sufficient that* $D(x_1, \ldots, x_n)$ *generate* $\mathscr{D}_{B/A}$.

Proof. Necessity has already been proved. Suppose that $d = D(x_1, \ldots, x_n)$ generates $\mathscr{D}_{B/A}$. Let $(e_1, \ldots, e_n)$ be a base of B over A. Put $d' = D(e_1, \ldots, e_n)$ and $x_i = \sum_{j=1}^{n} a_{ij}e_j$ with $a_{ij} \in A$, $1 \le i \le n$. Then $d = \det(a_{ij})^2 d'$. By hypothesis $Ad = \mathscr{D}_{B/A} = Ad'$. Therefore, there exists $b \in A$ such that $d' = bd$. It follows that $d(1 - b\det(a_{ij})^2) = 0$. We know that d is not a divisor of zero, since otherwise every element of $Ad = \mathscr{D}_{B/A}$ would be a divisor of zero. Thus $1 - b\det(a_{ij})^2 = 0$. This means that $\det(a_{ij})$ is invertible; the matrix (a_{ij}) must be invertible, too. Consequently, $(x_1, \ldots, x_n)$ is a base of B over A.

Proposition 3. Let K *be a field which is finite or of characteristic zero, let* L *be an extension of finite degree n of* K, *and let* $\sigma_1, \ldots, \sigma_n$ *be the n distinct* K*-isomorphisms of* L *into an algebraically closed field* C *containing* K (§ 4, *Theorem* 1). *Then, if* $(x_1, \ldots, x_n)$ *is a base for* L *over* K,

$$D(x_1, \ldots, x_n) = \det(\sigma_i(x_j))^2 \neq 0. \tag{3}$$

Proof. The first equality follows from a simple calculation:

$$\begin{aligned} D(x_1, \ldots, x_n) &= \det(\mathrm{Tr}(x_i x_j)) = \det\Big(\sum_k \sigma_k(x_i x_j)\Big) = \det\Big(\sum_k \sigma_k(x_i)\sigma_k(x_j)\Big) \\ &= \det(\sigma_k(x_i))\cdot\det(\sigma_k(x_i)) = \det(\sigma_i(x_j))^2. \end{aligned}$$

It remains to show that $\det(\sigma_i(x_j)) \neq 0$. We look for a contradiction. If $\det(\sigma_i(x_j)) = 0$, there exist $u_1, \ldots, u_n \in C$, not all zero, such that $\sum_{i=1}^{n} u_i\sigma_i(x_j) = 0$ for all j. By linearity we conclude that $\sum_{i=1}^{n} u_i\sigma_i(x) = 0$ for all $x \in L$. This contradicts the following:

Lemma of Dedekind. Let G *be a group,* C *a field, and let* $\sigma_1, \ldots, \sigma_n$ *be distinct homomorphisms of* G *into the multiplicative group* C*. *Then the* σ_i*'s are linearly independent over* C (*i.e.* $\sum u_i\sigma_i(g) = 0$ *for all* $g \in G$ *implies that all the* u_i*'s are zero*).

Proof. If the σ_i's are linearly dependent, consider a non-trivial relation $\sum_i u_i\sigma_i = 0$ $(u_i \in C)$ such that the number q of the u_i's which are non-zero is minimum. After renumbering, we may suppose that

$$u_1\sigma_1(g) + \cdots + u_q\sigma_q(g) = 0 \quad \text{for all } g \in G. \tag{4}$$

We have $q \ge 2$, since the σ_i's are not zero. For g and h arbitrary in G, we see that

$$u_1\sigma_1(hg) + \cdots + u_q\sigma_q(hg) = u_1\sigma_1(h)\sigma_1(g) + \cdots + u_q\sigma_q(h)\sigma_q(g) = 0.$$

Multiply (4) by $\sigma_1(h)$ and subtract. It follows that

$$u_2(\sigma_1(h) - \sigma_2(h))\sigma_2(g) + \cdots + u_q(\sigma_1(h) - \sigma_q(h))\sigma_q(g) = 0.$$

As this holds for any $g \in G$ and as q has been chosen as small as possible, it follows that $u_2(\sigma_1(h) - \sigma_2(h)) = 0$. Thus $\sigma_1(h) = \sigma_2(h)$ for all $h \in G$, since $u_2 \neq 0$. But this contradicts the hypothesis that the σ_i's are distinct. Q.E.D.

Remark. Under the conditions of Proposition 3, the relation $D(x_1, \ldots, x_n) \neq 0$ means that the bilinear form $(x, y) \mapsto \mathrm{Tr}_{L/K}(xy)$ is *non-degenerate.* i.e. $\mathrm{Tr}_{L/K}(xy) = 0$ for all $y \in L$ implies $x = 0$. Thus the K-linear mapping which attaches to each $x \in L$ the K-linear form $s_x : y \mapsto \mathrm{Tr}_{L/K}(xy)$ is an injection of L in its dual $\mathrm{Hom}_K(L, K)$ (for the structure of vector space over K). As L and $\mathrm{Hom}_K(L, K)$ are of the same finite dimension n over K, it follows that $x \mapsto s_x$ is a bijection. The existence of "*dual bases*" of a vector space and its dual implies that, for any base $(x_1, \ldots, x_n)$ of L over K, there exists a base $(y_1, \ldots, y_n)$ such that

$$\text{(5)} \qquad \mathrm{Tr}_{L/K}(x_i y_j) = \delta_{ij} \quad (1 \leq i, \ j \leq n).$$

This remark will prove useful.

Theorem 1. Let A be an integrally closed ring, let K be its field of fractions, L an extension of finite degree n of K, and A′ the integral closure of A in L. Suppose K is of characteristic 0. Then A′ is an A-submodule of a free A-module of rank n.

Proof. Let $(x_1, \ldots, x_n)$ be a base of L over K. Each x_i is algebraic over K, so, for any i, we have an equation of the form $a_n x_i^n + a_{n-1} x_i^{n-1} + \cdots + a_0 = 0$ ($a_j \in A$ for all j). We may assume $a_n \neq 0$. Multiplying through by a_n^{n-1}, we see that $a_n x_i$ is integral over A. Put $x_i' = a_n x_i$. Then $(x_1', \ldots, x_n')$ is a base for L over K contained in A′.

According to the remark preceding this theorem, there is another base $(y_1, \ldots, y_n)$ of L over K such that $\mathrm{Tr}(x_i' y_j) = \delta_{ij}$((5)). Let $z \in A'$. Since $(y_1, \ldots, y_n)$ is a base for L over K, we may write $z = \sum_{j=1}^{n} b_j y_j$ with $b_j \in K$. For any i we have $x_i' z \in A'$ (since $x_i' \in A'$). Therefore, $\mathrm{Tr}(x_i' z) \in A$ (§ 6, Corollary of Proposition 2). Thus, $\mathrm{Tr}(x_i' z) = \mathrm{Tr}(\sum_j b_j x_i' y_j) = \sum_j b_j \mathrm{Tr}(x_i' y_j) = \sum b_j \delta_{ij} = b_i$. We may conclude that $b_i \in A$ for all i, which implies that A′ is a submodule of the free A-module $\sum_{j=1}^{n} Ay_j$. Q.E.D.

Corollary. Add to the hypotheses of Theorem 1 the assumption that A is principal. Then A′ is a free A-module of rank n.

Proof. A submodule of a free A-module is, under our additional assumption, free (Chapter I, § 5, Theorem 1, (b)) and of rank $\leq n$. On the other hand we have seen in the course of the proof of Theorem 1 that A′ contains a base of L over K. Therefore, it is of rank n. Q.E.D.

As an exercise, the reader who lacks familiarity with the remark preceding Theorem 1 should look for a more computational proof of this theorem: with the notations defined above, set $d = D(x_1', \ldots, x_n')$ and show that, if $z = \sum_i c_i x_i'$ ($c_i \in K$) is integral over A, then $dc_i \in A$ (calculate $\mathrm{Tr}(zx_j')$ and use Cramer's rule).

AN EXAMPLE OF THE CALCULATION OF A DISCRIMINANT

Let K be a field which is finite or of characteristic zero, let $L = K[x]$ be an extension of finite degree n of K, and let $F(X)$ be the minimal polynomial of x over K. Then

$$\text{(6)} \qquad D(1, x, \ldots, x^{n-1}) = (-1)^{\frac{1}{2}n(n-1)} N_{L/K}(F'(x))$$

(where $F'(X)$ denotes the derivative of $F(X)$). Denote by $x_1, \ldots, x_n$ the roots of $F(X)$ in an extension of K; they are conjugates of x (§ 3, Proposition 3, and § 4). We see that

$$\begin{aligned} D(1, x, \ldots, x^{n-1}) &= \det(\sigma_i(x^j))^2 \quad \text{(by Proposition 3)} \\ &= \det(x_i^j)^2 = [\prod_{i<j} (x_i - x_j)]^2 \quad \text{(Vandermonde)} \\ &= c \prod_{i \neq j} (x_i - x_j) \quad (\text{where } c = (-1)^{\frac{1}{2}n(n-1)}) \\ &= c \prod_i (\prod_{j \neq i} (x_i - x_j)) \\ &= c \prod_i F'(x_i) = c N_{L/K}(F'(x)) \end{aligned}$$

(for the $F'(x_i)$'s are the conjugates of $F'(x)$).

In particular, applying (6) to the case where $F(X)$ is a trinomial $X^n + aX + b$ (a and $b \in K$) and putting $y = F'(x)$, we obtain

$$y = nx^{n-1} + a = -(n-1)a - nbx^{-1}$$

(since $x^n + ax + b = 0$ implies that $nx^{n-1} = -na - nbx^{-1}$). We obtain from this $x = -nb(y + (n-1)a)^{-1}$. The minimal polynomial of y over K is the numerator of $b^{-1} F(-nb(Y + (n-1)a)^{-1})$; the result of the computation is

$$(Y + (n-1)a)^n - na(Y + (n-1)a)^{n-1} + (-1)^n b^{n-1}.$$

The norm of y is $(-1)^n$ times the constant term of this polynomial, i.e.

$$n^n b^{n-1} + (-1)^{n-1}(n-1)^{n-1} a^n.$$

Thus,

$$\text{(7)} \qquad D(1, x, \ldots, x^{n-1}) = (-1)^{\frac{1}{2}n(n-1)}(n^n b^{n-1} + (-1)^{n-1}(n-1)^{n-1} a^n).$$

For $n = 2$ (respectively, 3) we rediscover the well-known expressions $a^2 - 4b$ (respectively, $-27b^2 - 4a^3$).

2.8. The terminology of number fields

We call any finite (and therefore algebraic) extension of **Q** an *algebraic number field* (or number field). For a number field K, the degree $[K : \mathbf{Q}]$ is called the *degree* of K. A number field of degree 2 (respectively, 3) is called a *quadratic field* (cf. § 5) (respectively, *cubic field*). A number field always has characteristic 0.

The elements of a number field K which are integral over **Z** are called the *integers* of K. They form a subring A of K (§ 1, Corollary 2 of Proposition 1). This ring A is a *free* **Z**-module of rank $[K : \mathbf{Q}]$ (§ 7, corollary of Theorem 1). The discriminants of the bases of the **Z**-module A differ by a unit in **Z** (§ 7, Definition 2), a unit which is even a square in **Z** (§ 7, Proposition 1). This can only be $+1$, i.e. the discriminant of the

Z-module A is a well-defined element of **Z**. It is called the *absolute discriminant* or the *discriminant* of K.

We frequently, by abuse of language, attribute to K notions which are defined relative to A. Thus when we speak of ideals (or units) of K, we mean ideals (or units) of A.

2.9. Cyclotomic fields

We call any number field generated over $\mathbf{Q}$ by roots of unity a *cyclotomic field*. Given a prime number p, we write z for a primitive pth root of unity (in $\mathbf{C}$ for example). We are going to study the cyclotomic field $\mathbf{Q}[z]$. The number z is a root of the polynomial $X^p - 1$. Since $z \neq 1$, it is also a root of the polynomial $(X^p - 1)/(X - 1) = X^{p-1} + X^{p-2} + \cdots + X + 1$, which is called a *cyclotomic polynomial*. It is not obvious that this polynomial is irreducible over $\mathbf{Q}$ (i.e. that the field $\mathbf{Q}[z]$ is of degree $p - 1$). In order to prove that this is indeed the case we need:

Eisenstein's irreducibility criterion. Let A *be a principal ideal ring, p a prime element of* A, *and*

$$F(X) = X^n + a_{n-1}X^{n-1} + \cdots + a_1X + a_0 \in A[X]$$

such that p divides $a_i (0 \leq i \leq n - 1)$ and p^2 does not divide a_0. Then F(X) *is irreducible over the field of fractions of* A.

Proof. Suppose that $F = G \cdot H$ with G and $H \in K[X]$, both G and H monic polynomials. The roots of F are integral over A. Any root of G or H is a root of F, therefore also integral over A. The coefficients of G (resp. H) are sums of products of roots of G (resp. H); they are therefore also integral over A (§ 1, Corollary 1 of Proposition 1). Since A is principal, it is integrally closed (§ 2, Example 2). Therefore $G \in A[X]$ and $H \in A[X]$.

Now let $\overline{F}$, $\overline{G}$, and $\overline{H}$ be the images of F, G, and H in $(A/Ap)[X]$, so $\overline{F} = \overline{G} \cdot \overline{H}$. According to the hypothesis on the a_i's we have $\overline{F} = X^n$. Since A/Ap is an integral domain, the factorization $X^n = \overline{G} \cdot \overline{H}$ is necessarily of the form $X^n = X^q \cdot X^{n-q}$ (since $\overline{G}$ and $\overline{H}$ are monic), thus $\overline{G} = X^q$ and $\overline{H} = X^{n-q}$. If G and H are both non-constant, then p divides the constant terms of both G and H. Therefore p^2 divides the constant term a_0 of F, and this contradicts the hypothesis. Thus, either G or H is constant, and F is irreducible. Q.E.D.

Example. The polynomial $X^3 - 2X + 6$ is irreducible over $\mathbf{Q}$ (take $p = 2$, $A = \mathbf{Z}$).

Theorem 1. For any prime number p the cyclotomic polynomial $X^{p-1} + X^{p-2} + \cdots + X + 1$ *is irreducible in* $\mathbf{Q}[X]$.

Proof. Put $X = Y + 1$. Then

$$X^{p-1} + X^{p-2} + \cdots + X + 1 = \frac{X^p - 1}{X - 1} = \frac{(Y + 1)^p - 1}{Y}$$

$$= Y^{p-1} + \sum_{j=p-1}^{1} \binom{p}{j} Y^{j-1} = F_1(Y).$$

If $F_1(Y)$ is irreducible, then so is the cyclotomic polynomial. Observing that p divides each of the binomial coefficients $\binom{p}{j}$ and that p^2 does not divide the constant term $\binom{p}{1} = p$, we conclude that $F_1(Y)$ is irreducible (Eisenstein's criterion). Q.E.D.

Theorem 1 implies that $\mathbf{Q}[z]$ is of degree $p - 1$. Therefore $(1, z, \ldots, z^{p-2})$ is a base for $\mathbf{Q}[z]$ over $\mathbf{Q}$. We are going to study the ring of integers of $\mathbf{Q}[z]$ and show that it is $\mathbf{Z}[z]$.

For this purpose we need to calculate some traces and norms (we write $\mathrm{Tr}(x)$ and $\mathrm{N}(x)$ in place of $\mathrm{Tr}_{\mathbf{Q}(z)/\mathbf{Q}}(x)$ and $\mathrm{N}_{\mathbf{Q}(z)/\mathbf{Q}}(x)$). Let us note that the conjugates of z over $\mathbf{Q}$ are the z^j's $(j = 1, \ldots, p-1)$ (Theorem 1).

The irreducibility of the cyclotomic polynomial implies immediately:

$$\mathrm{Tr}(z) = -1 \quad \text{and} \quad \mathrm{Tr}(1) = p - 1. \tag{1}$$

Therefore, $\mathrm{Tr}(z^j) = -1$ for $j = 1, \ldots, p-1$, and thus

$$\mathrm{Tr}(1 - z) = \mathrm{Tr}(1 - z^2) = \cdots = \mathrm{Tr}(1 - z^{p-1}) = p. \tag{2}$$

On the other hand, the calculation done in Theorem 1 shows that $\mathrm{N}(z - 1) = (-1)^{p-1}p$, from which it follows that $\mathrm{N}(1 - z) = p$. As the norm of $(1 - z)$ is the product of the conjugates of $1 - z$, we have

$$p = (1 - z)(1 - z^2)\ldots(1 - z^{p-1}). \tag{3}$$

Let us write A for the ring of integers in $\mathbf{Q}[z]$. Evidently a contains z and its powers. We are going to show that

$$\mathrm{A}\cdot(1 - z) \cap \mathbf{Z} = p\cdot\mathbf{Z}. \tag{4}$$

We know that $p \in \mathrm{A}\cdot(1 - z)$, (formula (3)). Thus, $\mathrm{A}\cdot(1 - z) \cap \mathbf{Z} \supset p\cdot\mathbf{Z}$. Since $p\cdot\mathbf{Z}$ is a maximal ideal of $\mathbf{Z}$, the relation $\mathrm{A}\cdot(1 - z) \cap \mathbf{Z} \neq p\cdot\mathbf{Z}$ implies $\mathrm{A}\cdot(1 - z) \cap \mathbf{Z} = \mathbf{Z}$, i.e. that $1 - z$ is a unit in A. But in this case the conjugates $(1 - z^j)$ of $1 - z$ must also be units; p must be a unit in $\mathrm{A} \cap \mathbf{Z}$ by (3); and thus p^{-1} must belong to $\mathbf{Z}$, which is absurd (§ 2, Example 2).

Let us show that, for any $y \in \mathrm{A}$,

$$\mathrm{Tr}(y(1 - z)) \in p\cdot\mathbf{Z}. \tag{5}$$

Each conjugate $y_j(1 - z^j)$ of $y(1 - z)$ is a multiple (in A) of $1 - z^j$, which is itself a multiple of $1 - z$, since $1 - z^j = (1 - z)(1 + z + \cdots + z^{j-1})$. Since the trace is the sum of the conjugates, we have $\mathrm{Tr}(y(1 - z)) \in \mathrm{A}\cdot(1 - z)$. (5) now follows immediately from (4), for the trace of an integer belongs to $\mathbf{Z}$ (§ 6, corollary of Proposition 2).

Now we are ready to determine the ring of integers of $\mathbf{Q}[z]$.

Theorem 2. Let p be a prime number and z a primitive pth root of unity in $\mathbf{C}$. *Then the ring* A *of integers of the cyclotomic field* $\mathbf{Q}[z]$ *is* $\mathbf{Z}[z]$, *and* $(1, z, \ldots, z^{p-2})$ *is a base of the* $\mathbf{Z}$*-module* A.

Proof. Let $x = a_0 + a_1 z + \cdots + a_{p-2}z^{p-2}$ $(a_i \in \mathbf{Q})$ be an element of A. Then

$$x(1 - z) = a_0(1 - z) + a_1(z - z^2) + \cdots + a_{p-2}(z^{p-2} - z^{p-1}).$$

Taking traces and making use of (1) and (2), we obtain $\mathrm{Tr}(x(1-z)) = a_0\mathrm{Tr}(1-z) = a_0 p$. By (5) $pa_0 \in p\mathbf{Z}$, so $a_0 \in \mathbf{Z}$. Since $z^{-1} = z^{p-1}$, $z^{-1} \in A$, therefore,

$$(x - a_0)z^{-1} = a_1 + a_2 z + \cdots + a_{p-2}z^{p-3} \in A.$$

By the same argument as before, $a_1 \in \mathbf{Z}$. Applying the same argument successively, we conclude that each $a_i \in \mathbf{Z}$. Q.E.D.

Remark. The results of this section easily extend to the case of cyclotomic fields $\mathbf{Q}[t]$, where t is a primitive p^rth root of unity (p prime). Such a field is of degree $p^{r-1}(p-1)$, and its ring of integers is $\mathbf{Z}[t]$. The minimal polynomial of t over $\mathbf{Q}$ is

$$X^{p^{r-1}(p-1)} + X^{p^{r-1}(p-2)} + \cdots + X^{p^{r-1}} + 1 = \frac{X^{p^r} - 1}{X^{p^{r-1}} - 1}.$$

APPENDIX

The field C of complex numbers is algebraically closed

Let K be a field and consider the following statements:

(a) Any polynomial of positive degree over K is a product of polynomials of degree one (linear polynomials).

(b) Any polynomial of positive degree has a root in K.

Clearly (a) implies (b). Conversely, if (b) is true, if P(X) is a polynomial of degree $d \geq 1$ over K, and if $a \in K$ is a root of P(X), then P(X) is a multiple of $X - a$, and induction on the degree d of P(X) shows that (a) is true. A field K which satisfies the equivalent conditions (a) and (b) is called *algebraically closed.*

We are going to show, by a method essentially due to Lagrange, that $\mathbf{C}(=\mathbf{R}[i], i^2 = -1)$ is algebraically closed. We shall make use of only the following facts:

1. Any polynomial of odd degree over $\mathbf{R}$ has a root in $\mathbf{R}$; this is a special case of Weierstrass's theorem on intermediate values.
2. Any quadratic polynomial over $\mathbf{C}$ has its roots in $\mathbf{C}$. The elementary "formula" for a root of $aX^2 + bX + c = 0$ reduces the question to that of showing that any $z = a + ib \in \mathbf{C}$ $(a,b \in \mathbf{R})$ has a square root in $\mathbf{C}$. But $(x + iy)^2 = a + ib$ $(x, y \in \mathbf{R})$ is equivalent to $x^2 - y^2 = a$ and $2xy = b$; it follows that $a^2 + b^2 = (x^2 + y^2)^2$ and $x^2 + y^2 = \sqrt{a^2 + b^2}$. Clearly, one may find x^2 and y^2, hence x and y, satisfying these equations.
3. Given a non-constant polynomial $P(X) \in K[X]$, there exists an extension K′ of K such that P(X) factors into a product of linear polynomials in K′[X]. There was an easy proof of this fact in Proposition 3 of § 3 (a proof almost independent of other results in this chapter; it suffices to know that, if F(X) is irreducible, K[X]/(F(X)) is a field. Then, use an induction).
4. The relations between coefficients and roots of a polynomial.
5. The fact that a symmetric polynomial $G(X_1, \ldots, X_n) \in K[X_1, \ldots, X_n]$ is a polynomial in the elementary symmetric functions $\sum X_i, \sum X_i X_j, \ldots, X_1 \ldots X_n$ of the X_i's.

Theorem. The field **C** *of complex numbers is algebraically closed.*

Proof. We shall prove property (b), that any non-constant polynomial has a root in **C**. Observing that $F(X) = P(X)\bar{P}(X)$ ($\bar{P}$: the polynomial whose coefficients are the complex conjugates of the coefficients of P) has roots in **C** if and only if P does, we see that we need consider only polynomials with real coefficients.

Now we write the degree of F(X) ($\in \mathbf{R}[X]$) in the form $d = 2^n q$ where q is odd. We argue by induction on the exponent n of 2. For $n = 0$, d is odd and F(X) has a root in **R** (cf. (1)). Suppose $n \geq 1$. By (3) there exists an extension K′ of **C** and $x_1, \ldots, x_d \in K'$ such that $F(X) = \prod_{i=1}^{d} (X - x_i)$ (assuming, without loss of generality, that F(X) is monic).

Let c be an arbitrary element of **R** and consider the elements $y_{ij} = x_i + x_j + cx_ix_j$ of $K'(i \leq j)$. The cardinality of $(y_{ij}), 1 \leq i \leq j \leq d$, is $\frac{1}{2}d(d+1) = 2^{n-1}q(d+1)$, where $q(d+1)$ is odd. The polynomial $G(X) = \prod_{i \leq j} (X - y_{ij})$ has as coefficients symmetric polynomials in the x_i's with real coefficients. By (5) the coefficients are polynomials in the elementary symmetric functions of the x_i's; these polynomials themselves have real coefficients. Therefore, the coefficients of G(X) are real (by (4)). As its degree is of the form $2^{n-1}x$ (x odd), the induction hypothesis implies that it has a root $z_c \in \mathbf{C}$. One of the y_{ij}'s, say $y_{i(c), j(c)} = x_{i(c)} + y_{j(c)} + cx_{i(c)}x_{j(c)}$, is therefore equal to z_c.

Now, since **R** is *infinite* and since the set of pairs (i, j) $(i \leq j)$ is finite, there exist two distinct real numbers c and c' such that $i(c) = i(c')$ and $j(c) = j(c')$. Denote these indices by r and s, respectively. Then $x_r + x_s + cx_rx_s = z_c \in \mathbf{C}$ and $x_r + x_s + c'x_rx_s = z_{c'} \in \mathbf{C}$. Taking linear combinations, we may conclude that $x_r + x_s \in \mathbf{C}$ and $x_rx_s \in \mathbf{C}$. Therefore, by (4), x_r and x_s are roots of a quadratic equation with coefficients in **C**. We may conclude that $x_r, x_s \in \mathbf{C}$ (by (2)). Thus F(X) has a root in **C** and the theorem is proved.

Chapter III

Noetherian rings and Dedekind rings

We refer the reader who wonders why we discuss Dedekind rings to § 4, the example and the discussion following Theorem 1. Noetherian rings are more general than Dedekind rings. We define Noetherian rings and develop a few of their properties in order to place these properties in their natural context, as well as because Noetherian rings are of fundamental importance in other areas of algebra and in algebraic geometry. Finally, the generalization of certain results regarding Noetherian rings to the case of Noetherian modules is another example of "linearization", a technique whose power the reader has already observed.

3.1. Noetherian rings and modules

In Chapter I, § 4, we proved the following:

Theorem 1. Let A *be a ring and* M *an* A*-module. The following statements are equivalent.*

(a) *Every non-empty collection of submodules of* M *contains a maximal element.*
(b) *Every increasing sequence of submodules of* M *is stationary.*
(c) *Every submodule of* M *is of finite type.*

Definition 1. An A-module M is called Noetherian if it satisfies the equivalent conditions of Theorem 1. A ring A is called Noetherian if, considered as an A-module, it is a Noetherian module.

We have seen (Chapter I, § 4, corollary of Theorem 1) that a principal ideal ring is Noetherian.

Proposition 1. Let A *be a ring,* E *an* A*-module, and* E′ *a submodule of* E*. In order that* E *be Noetherian it is necessary and sufficient that* E′ *and* E/E′ *be Noetherian.*

Proof. First, we prove necessity. Suppose E is Noetherian. The lattice of submodules of E′ (respectively, E/E′) is isomorphic to the lattice of submodules of E contained in E′ (respectively, containing E′). Thus E′ and E/E′ are Noetherian by (a) or (b).

Conversely, suppose E′ and E/E′ are Noetherian. Let $(F_n)_{n \geq 0}$ be an increasing sequence of submodules of E. As E′ is Noetherian, there is an integer n_0 such that $F_n \cap E' = F_{n+1} \cap E'$ for all $n \geqslant n_0$. As E/E′ is Noetherian, there is an integer n_1 such that

$(F_n + E')/E' = (F_{n+1} + E')/E'$ for all $n \geq n_1$. Therefore, $F_n + E' = F_{n+1} + E'$ for $n \geq n_1$. Take $n \geq \sup(n_0, n_1)$. We shall show that $F_n = F_{n+1}$. It suffices to show that $F_{n+1} \subset F_n$. To see this take $x \in F_{n+1}$. Since $F_{n+1} + E' = F_n + E'$, there exists $y \in F_n$ and $z', z'' \in E'$ such that $x + z' = y + z''$. Thus, $x - y = z'' - z' \in F_{n+1} \cap E'$. Note that $F_{n+1} \cap E' = F_n \cap E'$. Thus, since $x - y$ and y belong to F_n, $x \in F_n$ too. We conclude that $F_{n+1} = F_n$ for all $n \geq \sup(n_0, n_1)$; thus E is Noetherian by (b). Q.E.D.

Corollary 1. Let A *be a ring and let* $E_1, \ldots, E_n$ *be Noetherian* A*-modules. Then the* A*-module product* $\prod_{i=1}^{n} E_i$ *is Noetherian.*

Proof. For $n = 2$ E_1 may be identified with the submodule $E_1 \times (0)$ of $E_1 \times E_2$, and the corresponding quotient module is isomorphic to E_2. Our assertion follows from Proposition 1. The general case is proved by induction on n.

Corollary 2. Let A *be a Noetherian ring and let* E *be an* A*-module of finite type. Then* E *is a Noetherian module (and, therefore, all its submodules are of finite type).*

Proof. By Chapter I, § 4, E is isomorphic to a quotient module A^n/R (n being the cardinality of a finite set of generators of E). Corollary 1 implies that A^n is Noetherian and this fact, combined with Proposition 1, implies that A^n/R is Noetherian too.

3.2. An application concerning integral elements

Proposition 1. Let A *be a Noetherian integrally closed ring. Let* K *be its field of fractions,* L *a finite extension of* K*, and* A′ *the integral closure of* A *in* L*. Suppose that* K *is of characteristic* 0*. Then* A′ *is an* A*-module of finite type and a Noetherian ring.*

Proof. We know that A′ is a submodule of a free A-module of rank n (Chapter II, § 7, Theorem 1). Thus A′ is an A-module of finite type (§ 1, Corollary 2 of Proposition 1), and, therefore, a Noetherian module (ibid.). On the other hand, the ideals of A′ are special cases of A-submodules of A′. They satisfy the maximal condition (§ 1, Theorem 1, (a)), so A′ is a Noetherian ring.

Example. The ring of integers of a number field is *Noetherian* (take $A = \mathbf{Z}$, $K = \mathbf{Q}$).

3.3. Some preliminaries concerning ideals

An ideal $\mathfrak{p}$ of a ring A is called *prime* if the quotient ring $A/\mathfrak{p}$ is an *integral domain* Equivalently, the relations $x \in A - \mathfrak{p}$, $y \in A - \mathfrak{p}$ entail $xy \in A - \mathfrak{p}$, i.e. $A - \mathfrak{p}$ is stable under multiplication.

In order that an ideal $\mathfrak{m}$ of A be *maximal* (i.e. maximal among the ideals of A distinct from A), it is necessary and sufficient that $A/\mathfrak{m}$ contain no ideals besides itself and (0), i.e. that $A/\mathfrak{m}$ be a *field*. Thus, *every maximal ideal is prime.* The converse is false, as the ideal (0) of $\mathbf{Z}$ is prime but not maximal.

Lemma 1. Let A *be a ring,* $\mathfrak{p}$ *a prime ideal of* A*, and let* A′ *be a subring of* A*. Then* $\mathfrak{p} \cap A'$ *is a prime ideal of* A′*.*

Proof. $\mathfrak{p} \cap A'$ is the kernel of the composition of the homomorphisms $A' \to A \to A/\mathfrak{p}$, so there is an injective homomorphism $A'/\mathfrak{p} \cap A' \to A/\mathfrak{p}$. Clearly, a subring of an integral domain is an integral domain. Q.E.D.

Given two ideals $\mathfrak{a}$ and $\mathfrak{b}$ of a ring A, we define the *product* of $\mathfrak{a}$ and $\mathfrak{b}$ not as the set of products ab where $a \in \mathfrak{a}$ and $b \in \mathfrak{b}$, but as the set of all *finite sums* $\sum a_i b_i$ of such products. One sees immediately that $\mathfrak{ab}$ is an ideal of A. We have:

$$\mathfrak{ab} \subset \mathfrak{a} \cap \mathfrak{b}. \tag{1}$$

The two expressions are not always equal. In a principal ideal ring the left-hand side corresponds to the product of ideal generators and the right-hand side to the least common multiple of generators.

Ideal multiplication is associative and commutative. A acts as an identity element in the monoid.

Given an A-module E, a submodule F, and an ideal $\mathfrak{a}$ of A, we define in the same way the product $\mathfrak{a}$F. It is a submodule of E.

Lemma 2. If a prime ideal $\mathfrak{p}$ of a ring A *contains a product $\mathfrak{a}_1 \mathfrak{a}_2 \ldots \mathfrak{a}_n$ of ideals, then $\mathfrak{p}$ contains at least one of the ideals $\mathfrak{a}_i$.*

Proof. If $\mathfrak{a}_i \not\subset \mathfrak{p}$ for any i, then there exists $a_i \in \mathfrak{a}_i - \mathfrak{p}$ for all i. Therefore, $a_1 \ldots a_n \notin \mathfrak{p}$, since $\mathfrak{p}$ is prime. But $a_1 \ldots a_n \in \mathfrak{a}_1 \ldots \mathfrak{a}_n$, which contradicts the hypothesis of the lemma. Q.E.D.

Lemma 3. In a Noetherian ring every ideal contains a product of prime ideals. In a Noetherian integral domain A, *every non-zero ideal contains a product of non-zero prime ideals.*

Proof. We are going to make use of a type of argument which occurs rather frequently in the theory of Noetherian rings. Let us prove the second assertion (the proof of the first is analogous; it suffices to delete the word "non-zero" three times). We look for a contradiction. Assume that the collection Φ of non-zero ideals of A which contain no product of non-zero prime ideals is not empty. Since A is Noetherian, Φ contains a maximal element $\mathfrak{b}$ (§ 1, Theorem 1, (a)). The ideal $\mathfrak{b}$ cannot be prime; otherwise $\mathfrak{b}$ would not belong to Φ. Thus, there exist $x, y \in A - \mathfrak{b}$ such that $xy \in \mathfrak{b}$. The ideals $\mathfrak{b} + Ax$ and $\mathfrak{b} + Ay$ contain $\mathfrak{b}$ as a proper subset. Therefore, since $\mathfrak{b}$ is maximal, they do not belong to Φ. It follows that they both contain products of non-zero prime ideals:

$$\mathfrak{b} + Ax \supset \mathfrak{p}_1 \ldots \mathfrak{p}_n \quad \text{and} \quad \mathfrak{b} + Ay \supset \mathfrak{q}_1 \ldots \mathfrak{q}_r.$$

Since $xy \in \mathfrak{b}$,

$$(\mathfrak{b} + Ax)(\mathfrak{b} + Ay) \subset \mathfrak{b},$$

whence $\mathfrak{p}_1 \ldots \mathfrak{p}_n \mathfrak{q}_1 \ldots \mathfrak{q}_r \subset \mathfrak{b}$. Q.E.D.

Now let A be an integral domain and let K be its field of fractions. We call any A-submodule I of K for which there exists $d \in A - (0)$ such that $d \cdot I \subset A$ a *fractional ideal* of A (or of K with respect to A). This means that the elements of I have a "common denominator" $d \in A$. The ordinary ideals of A are fractional ideals (with $d = 1$). We sometimes call them *integral ideals* to distinguish them from fractional ideals. Any A-submodule I of finite type contained in K is a fractional ideal. This follows from the fact that, if $(x_1, \ldots, x_n)$ is a finite set of generators for I, the x_i's have a common denominator d (e.g. the product of the denominators d_i, where $x_i = a_i d_i^{-1}$, with $a_i, d_i \in A$), and d is a common denominator for I. Conversely, if A is *Noetherian*, every fractional ideal I is an A-module of finite type, i.e. $I \subset d^{-1}A$ and $d^{-1}A$, being an A-module isomorphic to A, is a Noetherian module.

We define the *product* II′ of two fractional ideals I and I′ as the set of finite sums $\sum x_i y_i$ where $x_i \in I$ and $y_i \in I'$. If I and I′ are fractional ideals, with common denominators d and d', then the sets $I \cap I'$, $I + I'$, and II′ are all fractional *ideals*. They are clearly A-submodules of K and they have as common denominators d (or d'), dd', and dd', respectively. The non-zero fractional ideals of A constitute a commutative *monoid* under multiplication.

3.4. Dedekind rings

Definition 1. An integral domain A *is called a Dedekind ring if it is Noetherian and integrally closed, and if every non-zero prime ideal of* A *is maximal.*

The ring **Z**, and more generally any principal ideal ring, is a Dedekind ring. The following theorem implies that the ring of integers in a number field is a Dedekind ring.

Theorem 1. Let A *be a Dedekind ring,* K *its field of fractions,* L *an extension of finite degree of* K, *and* A′ *the integral closure of* A *in* L. *Assume* K *is of characteristic* 0. *Then* A′ *is a Dedekind ring and an* A*-module of finite type.*

Proof. The ring A′ is integrally closed by construction. It is Noetherian and an A-module of finite type by the proposition of § 2. It remains to show that every prime ideal $\mathfrak{p}' \neq (0)$ of A′ is maximal. For this purpose choose an element $x \in \mathfrak{p}' - (0)$ and consider an equation of integral dependence of x over A, the degree of which is a minimum:

$$x^n + a_{n-1}x^{n-1} + \cdots + a_1 x + a_0 = 0 \quad (a_i \in A). \tag{1}$$

Then $a_0 \neq 0$, since otherwise one could factor out an x and obtain an equation of lower degree. By (1), we have $a_0 \in A'x \cap A \subset \mathfrak{p}' \cap A$. Therefore, $\mathfrak{p}' \cap A \neq (0)$. Since $\mathfrak{p}' \cap A$ is a prime ideal of A (§ 3, Lemma 1), we see that $\mathfrak{p}' \cap A$ is a maximal ideal of A and $A/\mathfrak{p}' \cap A$ is a field. But $A/\mathfrak{p}' \cap A$ may be identified with a subring of $A'/\mathfrak{p}'$, and $A'/\mathfrak{p}'$ is *integral* over $A/\mathfrak{p}' \cap A$ (since A′ is integral over A). Thus $A'/\mathfrak{p}'$ is a field (Chapter II, § 1, Proposition 3), so $\mathfrak{p}'$ is maximal. Q.E.D.

Interest in Dedekind rings arises from the fact that the ring of integers in a number field is a Dedekind ring, but not always a principal ideal ring.

Example. Consider the ring of integers $A = \mathbf{Z}[\sqrt{-5}]$ in $\mathbf{Q}[\sqrt{-5}]$ (Chapter II, § 5, Theorem 1). Observe that

$$(1 + \sqrt{-5})(1 - \sqrt{-5}) = 2 \cdot 3. \tag{2}$$

The norms of the four factors are, respectively, 6, 6, 4, and 9. Note that $1 + \sqrt{-5}$ can have no non-trivial divisor in A, since the norm of such a divisor would have to be a non-trivial divisor of 6. This is impossible, because the equations $a^2 + 5b^2 = 2$ and $a^2 + 5b^2 = 3$ have no solutions in **Z**. If A were principal, the prime element

$1 + \sqrt{-5}$, which divides the product $2 \cdot 3$ by (2), would have to divide either 2 or 3. But then, taking norms, we see that 6 would divide 4 or 9, which is not the case.

Historically, the arithmetician Kummer (1810–1893) observed that the rings of integers in certain number fields were not principal ideal rings (in fact, certain cyclotomic fields; Kummer observed this in connection with his work on Fermat's equation, cf. I, § 2). In order, at least in part, to get around this inconvenience, he and Dedekind (1831–1916) introduced the notion of *ideal*. Dedekind then studied the rings which now carry his name. The most important property of principal ideal rings is unique factorization into products of primes. There is an elegant generalization of this property to the case of Dedekind rings. In a Dedekind ring ideals factor uniquely into products of prime ideals. There are many interesting consequences of this unique factorization, which we intend now to describe precisely and prove.

Theorem 2. Let A *be a Dedekind ring which is not a field. Every maximal ideal of* A *is invertible in the monoid of fractional ideals of* A.

Proof. Let $\mathfrak{m}$ be a maximal ideal of A. Then $\mathfrak{m} \neq (0)$, since A is not a field. Put

$$\mathfrak{m}' = \{x \in \mathrm{K} \mid x\mathfrak{m} \subset \mathrm{A}\}. \tag{3}$$

Clearly, $\mathfrak{m}'$ is an A-submodule of K; any non-zero element of $\mathfrak{m}$ serves as a common denominator for the elements of $\mathfrak{m}'$. Thus $\mathfrak{m}'$ is a fractional ideal of A. It suffices to show that $\mathfrak{m}\mathfrak{m}' = \mathrm{A}$. We see that (3) implies that $\mathfrak{m}\mathfrak{m}' \subset \mathrm{A}$; on the other hand, $\mathrm{A} \subset \mathfrak{m}'$ (since $\mathfrak{m}$ is an ideal), so $\mathfrak{m} = \mathrm{A}\mathfrak{m} \subset \mathfrak{m}'\mathfrak{m}$. As $\mathfrak{m}$ is maximal and $\mathfrak{m} \subset \mathfrak{m}'\mathfrak{m} \subset \mathrm{A}$, either $\mathfrak{m}'\mathfrak{m} = \mathrm{A}$ or $\mathfrak{m}'\mathfrak{m} = \mathfrak{m}$. It remains to show that $\mathfrak{m}'\mathfrak{m} = \mathfrak{m}$ is impossible.

Now, if $\mathfrak{m}'\mathfrak{m} = \mathfrak{m}$ and if $x \in \mathfrak{m}'$, then $x\mathfrak{m} \subset \mathfrak{m}$, $x^2\mathfrak{m} \subset x\mathfrak{m} \subset \mathfrak{m}$, and $x^n\mathfrak{m} \subset \mathfrak{m}$ for any $n \in \mathbf{N}$ by induction. Thus any non-zero element $d \in \mathfrak{m}$ is a common denominator for all the powers x^n of x, $n \in \mathbf{N}$. It follows that $\mathrm{A}[x]$ is a fractional ideal of A. Since A is Noetherian, $\mathrm{A}[x]$ is an A-module of finite type (§ 3, near the end), so x is *integral* over A (Chapter II, § 1, Theorem 1). But A is integrally closed; therefore, $x \in \mathrm{A}$; and consequently $\mathfrak{m}'\mathfrak{m} = \mathfrak{m}$ implies $\mathfrak{m}' = \mathrm{A}$. It remains to show that $\mathfrak{m}' = \mathrm{A}$ is impossible.

For this purpose take a non-zero element $a \in \mathfrak{m}$. The ideal $\mathrm{A}a$ contains a product $\mathfrak{p}_1\mathfrak{p}_2 \ldots \mathfrak{p}_n$ of non-zero prime ideals (§ 3, Lemma 3). We may take n as small as possible. We have $\mathfrak{m} \supset \mathrm{A}a \supset \mathfrak{p}_1 \ldots \mathfrak{p}_n$, which means that $\mathfrak{m} \supset \mathfrak{p}_i$ for some i (§ 3, Lemma 2), say $i = 1$. As $\mathfrak{p}_1$ is maximal by hypothesis, $\mathfrak{m} = \mathfrak{p}_1$. Put $\mathfrak{b} = \mathfrak{p}_2 \ldots \mathfrak{p}_n$. Then $\mathrm{A}a \supset \mathfrak{m}\mathfrak{b}$ and $\mathrm{A}a \not\supset \mathfrak{b}$, since n was as small as possible. There thus exists $b \in \mathfrak{b}$ such that $b \notin \mathrm{A}a$. Since $\mathfrak{m}\mathfrak{b} \subset \mathrm{A}a$, $\mathfrak{m}b \subset \mathrm{A}a$, whence $\mathfrak{m}ba^{-1} \subset \mathrm{A}$. According to the definition (3) of $\mathfrak{m}'$, this means that $ba^{-1} \in \mathfrak{m}'$. But, since $b \notin \mathrm{A}a$, $ba^{-1} \notin \mathrm{A}$. Thus $\mathfrak{m}' \neq \mathrm{A}$. Q.E.D.

Theorem 3. Let A *be a Dedekind ring and let* P *be the set of non-zero prime ideals of* A. *Then*

(a) *Every non-zero fractional ideal* $\mathfrak{b}$ *of* A *may be uniquely expressed in the form*

$$\mathfrak{b} = \prod_{\mathfrak{p} \in \mathrm{P}} \mathfrak{p}^{n_{\mathfrak{p}}(\mathfrak{b})}, \tag{4}$$

where, for any $\mathfrak{p} \in \mathrm{P}$, $n_{\mathfrak{p}}(\mathfrak{b}) \in \mathbf{Z}$ *and, for almost all* $\mathfrak{p} \in \mathrm{P}$, $n_{\mathfrak{p}}(\mathfrak{b}) = 0$.

(b) *The monoid of non-zero fractional ideals of* A *is a group.*

Proof. First we prove the existence of (a), i.e. that any fractional ideal $\mathfrak{b}$ is a product of powers (≥ 0 or ≤ 0) of prime ideals. There exists $d \in A - (0)$ such that $d\mathfrak{b} \subset A$, i.e. such that $d\mathfrak{b}$ is an integral ideal of A, $\mathfrak{b} = (d\mathfrak{b}) \cdot (Ad)^{-1}$. We may, without loss of generality, prove (a) for integral ideals. Proceeding as in Lemma 3 of § 3, we consider the collection Φ of non-zero ideals in A which are not products of prime ideals. Suppose that Φ is not empty. Let $\mathfrak{a}$ be a maximal element of Φ (A is Noetherian). Then $\mathfrak{a} \neq A$, since A is the product of the empty collection of prime ideals. So $\mathfrak{a}$ is contained in a maximal ideal $\mathfrak{p}$, which is thus a maximal element in the collection of non-trivial ideals of A which contain $\mathfrak{a}$. Let $\mathfrak{p}'$ be the inverse fractional ideal of $\mathfrak{p}$. Since $\mathfrak{a} \subset \mathfrak{p}$, $\mathfrak{a}\mathfrak{p}' \subset \mathfrak{p}\mathfrak{p}' = A$. As $\mathfrak{p}' \supset A$, $\mathfrak{a}\mathfrak{p}' \supset \mathfrak{a}$; in fact $\mathfrak{a}\mathfrak{p}' \neq \mathfrak{a}$ (if $\mathfrak{a}\mathfrak{p}' = \mathfrak{a}$ and if $x \in \mathfrak{p}'$, then $x\mathfrak{a} \subset \mathfrak{a}$, $x^n\mathfrak{a} \subset \mathfrak{a}$ for all n, x integral over A, and $x \in A$ (as in Theorem 2). But this is impossible, since $\mathfrak{p}' \neq A$ (otherwise $\mathfrak{p}' = A$ and $\mathfrak{p}\mathfrak{p}' = \mathfrak{p}$).). According to the maximality of $\mathfrak{a}$ in Φ, we have $\mathfrak{a}\mathfrak{p}' \notin \Phi$, so $\mathfrak{a}\mathfrak{p}' = \mathfrak{p}_1 \ldots \mathfrak{p}_n$, a product of prime ideals. Multiplying by $\mathfrak{p}$, we see that $\mathfrak{a} = \mathfrak{p}\mathfrak{p}_1 \ldots \mathfrak{p}_n$. Thus every integral ideal of A is a product of prime ideals.

Let us consider next the uniqueness of (a). Suppose that

$$\prod_{\mathfrak{p} \in P} \mathfrak{p}^{n(\mathfrak{p})} = \prod_{\mathfrak{p} \in P} \mathfrak{p}^{m(\mathfrak{p})}, \quad \text{i.e.} \quad \prod_{\mathfrak{p} \in P} \mathfrak{p}^{n(\mathfrak{p}) - m(\mathfrak{p})} = A.$$

If $n(\mathfrak{p}) - m(\mathfrak{p}) \neq 0$ for some prime ideals $\mathfrak{p} \in P$, we may separate the positive and negative exponents and write:

(5)
$$\mathfrak{p}_1^{\alpha_1} \ldots \mathfrak{p}_r^{\alpha_r} = \mathfrak{q}_1^{\beta_1} \ldots \mathfrak{q}_s^{\beta_s},$$

where $\mathfrak{p}_i, \mathfrak{q}_j \in P$, $\alpha_i > 0$, $\beta_j > 0$, $\mathfrak{p}_i \neq \mathfrak{q}_j$ for all i and j. Thus $\mathfrak{p}_1$ contains $\mathfrak{q}_1^{\beta_1} \ldots \mathfrak{q}_s^{\beta_s}$; $\mathfrak{p}_1 \supset \mathfrak{q}_j$ for some j (§ 3, Lemma 2), say $\mathfrak{p}_1 \supset \mathfrak{q}_1$. But $\mathfrak{p}_1$ and $\mathfrak{q}_1$ are both maximal, which implies $\mathfrak{p}_1 = \mathfrak{q}_1$, which is a contradiction.

Finally (4) implies that $\prod_{\mathfrak{p} \in P} \mathfrak{p}^{-n_{\mathfrak{p}}(\mathfrak{b})}$ is the inverse of $\mathfrak{b}$ and this proves (b). Q.E.D.

Remark. We have just seen that the monoid I(A) of non-zero fractional ideals of a Dedekind ring is a group. The principal fractional ideals (i.e. those of the form Ax, $x \in K^*$) form a subgroup F(A) of I(A) (since $(Ax)(Ay)^{-1} = Axy^{-1}$). The quotient group $C(A) = I(A)/F(A)$ is called the *ideal class group* of A. In order that A be a principal ideal ring, it is necessary and sufficient that C(A) consist of a single element.

Let us complete this section with some formulas, in which $n_{\mathfrak{p}}(\mathfrak{b})$ denotes the exponent of $\mathfrak{p}$ in the factorization of $\mathfrak{b}$ into a product of prime ideals (cf. (4)).

(6)
$$n_{\mathfrak{p}}(\mathfrak{a}\mathfrak{b}) = n_{\mathfrak{p}}(\mathfrak{a}) + n_{\mathfrak{p}}(\mathfrak{b}) \quad \text{(proof obvious).}$$

(7)
$$\mathfrak{b} \subset A \Leftrightarrow n_{\mathfrak{p}}(\mathfrak{b}) \geq 0 \quad \text{for all } \mathfrak{p} \in P.$$

($\Rightarrow$ seen in the course of the proof of Theorem 3; $\Leftarrow$ obvious).

(8)
$$\mathfrak{a} \subset \mathfrak{b} \Leftrightarrow n_{\mathfrak{p}}(\mathfrak{a}) \geq n_{\mathfrak{p}}(\mathfrak{b}) \quad \text{for all } \mathfrak{p} \in P.$$

($\mathfrak{a} \subset \mathfrak{b}$ means the same as $\mathfrak{a}\mathfrak{b}^{-1} \subset A$. Now apply (6) and (7)).

(9)
$$n_{\mathfrak{p}}(\mathfrak{a} + \mathfrak{b}) = \inf(n_{\mathfrak{p}}(\mathfrak{a}), n_{\mathfrak{p}}(\mathfrak{b}))$$

($\mathfrak{a} + \mathfrak{b}$ is the least upper bound of $\mathfrak{a}$ and $\mathfrak{b}$ for ideal inclusion; to complete the proof use (8)).

$$n_{\mathfrak{p}}(\mathfrak{a} \cap \mathfrak{b}) = \sup(n_{\mathfrak{p}}(\mathfrak{a}), n_{\mathfrak{p}}(\mathfrak{b})) \tag{10}$$

(analogous argument involving greatest lower bounds; again use (8)).

3.5 The norm of an ideal

In this section, K *denotes a number field,* n *its degree, and* A *the ring of integers of* K. *We write* $N(x)$ *in place of* $N_{K/\mathbf{Q}}(x)$.

Proposition 1. If x is a non-zero element of A, *then* $|N(x)| = \text{card}(A/Ax)$.

Note that, since $x \in A$, we have $N(x) \in \mathbf{Z}$ (Chapter II, §6, corollary of Proposition 2), so the preceding formula makes sense.

Proof. We know that A is a free $\mathbf{Z}$-module of rank n (Chapter II, § 8), and Ax is a $\mathbf{Z}$-submodule of A. It is also of rank n, since multiplication by x maps A to Ax isomorphically. According to Chapter I, § 5, Theorem 1, there exists a base $(e_1, \ldots, e_n)$ of the $\mathbf{Z}$-module A together with elements c_i of $\mathbf{N}$ such that $(c_1e_1, \ldots, c_ne_n)$ is a base of Ax. Furthermore, the abelian group A/Ax is isomorphic to the finite abelian group $\prod_{i=1}^{n} \mathbf{Z}/c_i\mathbf{Z}$, whose order is $c_1c_2 \ldots c_n$. Write u for the $\mathbf{Z}$-linear mapping of A on Ax defined by $u(e_i) = c_ie_i$ for $i = 1, \ldots, n$. We have $\det(u) = c_1 \ldots c_n$. On the other hand $(xe_1, \ldots, xe_n)$ is also a base for Ax. There is thus an automorphism v of the $\mathbf{Z}$-module Ax such that $v(c_ie_i) = xe_i$. Then $\det(v)$ is invertible in $\mathbf{Z}$, so $\det(v) = \pm 1$. But $v \cdot u$ is multiplication by x, and its determinant is, by definition, $N(x)$ (Chapter II, § 6, Definition 1). Since $\det(v \cdot u) = \det(v) \cdot \det(u)$, we may conclude that $N(x) = \pm c_1 \ldots c_n = \pm \text{card}(A/Ax)$. Q.E.D.

Definition 1. Given a non-zero integral ideal $\mathfrak{a}$ *of* A, *we call the number* card(A/$\mathfrak{a}$) *the norm of* $\mathfrak{a}$ *and denote it by* N($\mathfrak{a}$).

Let us observe that N($\mathfrak{a}$) is *finite*. In fact, if a is a non-zero element of $\mathfrak{a}$, then A$a \subset \mathfrak{a}$, and A/$\mathfrak{a}$ may be identified with a quotient of A/Aa. Thus $\text{card}(A/\mathfrak{a}) \leq \text{card}(A/Aa)$, which is finite by Proposition 1. On the other hand we saw that, for a principal ideal Ab, $N(Ab) = |N(b)|$.

Proposition 2. If $\mathfrak{a}$ *and* $\mathfrak{b}$ *are both non-zero integral ideals of* A, *then* $N(\mathfrak{a}\mathfrak{b}) = N(\mathfrak{a})N(\mathfrak{b})$.

Proof. The ideal $\mathfrak{b}$ factors into a product of maximal ideals (§ 4, Theorem 3), and it suffices to show that $N(\mathfrak{a}\mathfrak{m}) = N(\mathfrak{a})N(\mathfrak{m})$ for $\mathfrak{m}$ maximal. Since $\mathfrak{a}\mathfrak{m} \subset \mathfrak{a}$, we have $\text{card}(A/\mathfrak{a}\mathfrak{m}) = \text{card}(A/\mathfrak{a})\ \text{card}(\mathfrak{a}/\mathfrak{a}\mathfrak{m})$. It thus suffices to show that $\text{card}(\mathfrak{a}/\mathfrak{a}\mathfrak{m}) = \text{card}(A/\mathfrak{m})$. Now $\mathfrak{a}/\mathfrak{a}\mathfrak{m}$ is an A-module annihilated by $\mathfrak{m}$, which means it may be considered as a vector space over A/$\mathfrak{m}$. Its subspaces are its A-submodules; and they are of the form $\mathfrak{q}/\mathfrak{a}\mathfrak{m}$, where $\mathfrak{q}$ is an ideal such that $\mathfrak{a}\mathfrak{m} \subset \mathfrak{q} \subset \mathfrak{a}$. But formula (8) of § 4 implies that there are no ideals between $\mathfrak{a}\mathfrak{m}$ and $\mathfrak{a}$. Therefore, the vector space $\mathfrak{a}/\mathfrak{a}\mathfrak{m}$ is of dimension one over A/$\mathfrak{m}$. This means that $\text{card}(\mathfrak{a}/\mathfrak{a}\mathfrak{m}) = \text{card}(A/\mathfrak{m})$. Q.E.D.

Chapter IV

Ideal classes and the unit theorem

The present chapter is devoted to two important finiteness theorems. For the proofs of these theorems we shall need some analysis. We proceed to develop this material first.

4.1. Preliminaries concerning discrete subgroups of $\mathbf{R}^n$

A subgroup H of $\mathbf{R}^n$ is discrete if and only if, for any compact subset K of $\mathbf{R}^n$, the intersection $H \cap K$ is finite. A typical example of a discrete subgroup of $\mathbf{R}^n$ is $\mathbf{Z}^n$. We are going to show that it is almost the only one.

Theorem 1. Let H *be a discrete subgroup of* $\mathbf{R}^n$. *Then* H *is generated (as a* **Z***-module) by* r *vectors which are linearly independent over* **R** *(so* $r \leq n$*).*

Proof. Let $(e_1, \ldots, e_r)$ be a set of elements of H which are linearly independent over **R**, where r is as large as possible (again, $r \leq n$). Let

$$P = \left\{x \in \mathbf{R}^n \;\middle|\; x = \sum_{i=1}^{r} \alpha_i e_i,\ 0 \leq \alpha_i \leq 1\right\}, \tag{1}$$

the parallelotope constructed on these vectors. Clearly, P is compact, so $P \cap H$ is finite. Take $x \in H$. From the maximality of the set $(e_1, \ldots, e_r)$ it follows that $x = \sum_{i=1}^{r} \lambda_i e_i$, $\lambda_i \in \mathbf{R}$. For $j \in \mathbf{Z}$ set

$$x_j = jx - \sum_{i=1}^{r} [j\lambda_i] e_i \tag{2}$$

(where $[\mu]$ denotes the largest integer less than or equal to $\mu \in \mathbf{R}$). Thus,

$$x_j = \sum_{i=1}^{r} (j\lambda_i - [j\lambda_i]) e_i,$$

from which it follows that $x_j \in P$ and, by (2), $x_j \in P \cap H$. If one notices that $x = x_1 + \sum_{i=1}^{r} [\lambda_i] e_i$, one sees that the **Z**-module H is generated by $P \cap H$ and is thus of *finite type*.

On the other hand, since $P \cap H$ is finite and $\mathbf{Z}$ is infinite, there exist distinct integers j and k such that $x_j = x_k$. It follows from (2) that $(j-k)\lambda_i = [j\lambda_i] - [k\lambda_i]$, which implies that the λ_i's are rational. Thus the $\mathbf{Z}$-module H is generated by a finite number of elements which are linear combinations with rational coefficients of the e_i's. Let d be a common denominator ($d \in \mathbf{Z} - (0)$) of these coefficients. Clearly, $dH \subset \sum_{i=1}^{r} \mathbf{Z}e_i$. Thus there exists a base (f_i) of the $\mathbf{Z}$-module $\sum_{i=1}^{r} \mathbf{Z}e_i$ and integers α_i such that $(\alpha_1 f_1, \ldots, \alpha_r f_r)$ generates dH (Chapter I, § 5, Theorem 1). Since the $\mathbf{Z}$-module dH has the same rank as H and since $H \supset \sum_{i=1}^{r} \mathbf{Z}e_i$, the rank of dH is $\geq r$. Therefore, the rank of dH equals r and the α_i's are non-zero. We may conclude that the f_i's are, like the e_i's, linearly independent over $\mathbf{R}$. The module dH, and consequently H itself, is generated (over $\mathbf{Z}$) by r vectors linearly independent over $\mathbf{R}$.

An application of the preceding. Let $t = (\theta_1, \ldots, \theta_n) \in \mathbf{R}^n$ such that at least one of the θ_i's is *irrational.* Write $(e_1, \ldots, e_n)$ for the canonical base of $\mathbf{R}^n$ and let H denote the subgroup of $\mathbf{R}^n$ generated by $(e_1, \ldots, e_n, t)$. The group H is not discrete; otherwise the methods employed in the proof of Theorem 1 would provide us with an expression for t as a rational linear combination of the e_i's, an absurdity. Therefore, for any $\epsilon > 0$ there exists a non-zero element of H whose distance from 0 is smaller than ϵ. Therefore, there exist integers $p_i \in \mathbf{Z}$, $q \in \mathbf{N}$, $q \neq 0$ such that $|q\theta_i - p_i| \leq \epsilon$, which means that

$$\left| \theta_i - \frac{p_i}{q} \right| \leq \frac{\epsilon}{q} \quad \text{for all } i = 1, \ldots, n.$$

Let us remark that, simply by picking the multiple n_i/q of $1/q$ nearest θ_i, we would obtain the cruder approximation

$$\left| \theta_i - \frac{n_i}{q} \right| \leq \frac{1}{2q} \; (n_i \in \mathbf{Z}), \quad \text{for any } q > 0.$$

The result proved in the last paragraph is a basic theorem in the very rich theory of approximation of irrational numbers by rationals. The reader interested in learning more about this subject should consult Koksma, "Diophantische Approximation", Berlin (Springer), 1936.

Definition 1. A discrete subgroup of rank n of $\mathbf{R}^n$ *is called a lattice in* $\mathbf{R}^n$.

By Theorem 1 a lattice is generated over $\mathbf{Z}$ by a base of $\mathbf{R}^n$, which is then a $\mathbf{Z}$-base for the given lattice. For each $\mathbf{Z}$-base $e = (e_1, \ldots, e_n)$ of a lattice H we shall write P_e for the half open parallelotope

$$P_e = \left\{ x \in \mathbf{R}^n \;\middle|\; x = \sum_{i=1}^{n} \alpha_i e_i, \text{ with } 0 \leq \alpha_i < 1 \right\}.$$

Thus every point of $\mathbf{R}^n$ is congruent modulo H to one and only one point of P_e for any fixed e (we say, in this case, that P_e is a *fundamental domain* for H). We shall write μ to denote the *Lebesgue measure* in $\mathbf{R}^n$, i.e. if S is a measurable subset of $\mathbf{R}^n$, $\mu(S)$ will stand for its measure (which we will also call its volume).

Lemma 1. The volume $\mu(P_e)$ is independent of the base e chosen for H.

Proof. Let $f = (f_1, \ldots, f_n)$ be another base of H. Then

$$f_i = \sum_{j=1}^{n} \alpha_{ij} e_j \quad \text{with } \alpha_{ij} \in \mathbf{Z}.$$

By calculus we know that $\mu(P_f) = |\det(\alpha_{ij})|\, \mu(P_e)$. The matrix (α_{ij}), being associated with a change of base, is invertible with an integer inverse matrix, so $\det(\alpha_{ij}) = \pm 1$. Thus, $\mu(P_f) = \mu(P_e)$. Q.E.D.

The volume of the parallelotope P_e associated with any base e of H is called the *volume of the lattice* H and is denoted $v(H)$ (the word "volume" is here an abuse of language since $\mu(H) = 0$; would it perhaps be better to speak of the "mesh" of the lattice H?).

Theorem 2 (Minkowski). *Let* H *be a lattice in* $\mathbf{R}^n$ *and let* S *be a measurable subset of* $\mathbf{R}^n$ *such that* $\mu(S) > v(H)$. *Then there exist two distinct points* $x, y \in S$ *such that* $x - y \in H$.

Proof. Let $e = (e_1, \ldots, e_n)$ be a **Z**-base of H and let P_e be the parallelotope associated with e. Since P_e is a fundamental domain for H, S is the disjoint union of subsets of the form $S \cap (h + P_e)$, $h \in H$. It follows that

$$\mu(S) = \sum_{h \in H} \mu(S \cap (h + P_e)). \tag{3}$$

Since μ is translation invariant,

$$\mu(S \cap (h + P_e)) = \mu((-h + S) \cap P_e).$$

The sets $(-h + S) \cap P_e (h \in H)$ cannot all be pairwise disjoint, otherwise $\mu(P_e) \geq \sum_{h \in H} \mu((-h + S) \cap P_e)$, which contradicts (3) and the hypothesis $\mu(P_e) = v(H) < \mu(S)$. Consequently, there exist two distinct elements h and h' of H such that $P_e \cap (-h + S) \cap (-h' + S) \neq \varphi$. Let x and y be elements of S such that $-h + x = -h' + y$. Then $x - y = h - h' \in H$ and $x \neq y$, since $h \neq h'$. Q.E.D.

Corollary. Let H *be a lattice in* $\mathbf{R}^n$ *and let* S *be a measurable subset of* $\mathbf{R}^n$ *which is symmetric with respect to* 0 *and convex. Assume that* S *satisfies at least one of the following two conditions:*

(a) $\mu(S) > 2^n v(H)$
or (b) $\mu(S) \geq 2^n v(H)$ *and* S *is compact.*

Then $S \cap (H - \{0\}) \neq \phi$.

Proof. In case (a) apply Theorem 1 to the set $S' = \frac{1}{2}S$ $(\mu(S') = 2^{-n}\mu(S) > v(H))$. Let y and z be distinct points of S' such that $y - z \in H$. Then $y - z$ also belongs to S, because $y - z = \frac{1}{2}(2y + (-2z))$ and S is both symmetric and convex. Therefore, $y - z \in S \cap (H - \{0\})$. To prove that case (b) also implies the conclusion of the corollary apply case (a) to $(1 + \epsilon)S$ for $\epsilon > 0$. Thus, $(H - \{0\}) \cap (1 + \epsilon)S$ is a non-empty compact set (it is even finite, since it is compact and discrete). Furthermore, $\bigcap_{\epsilon > 0} (H - \{0\}) \cap (1 + \epsilon)S \neq \phi$, since an intersection of non-empty, nested compact sets is never void. This means that there is a point of $H - \{0\}$ which belongs to $(1 + \epsilon)S$ for all $\epsilon > 0$; therefore, since S is compact, it belongs to S, too. Q.E.D.

The hypothesis of compactness is needed in (b). Consider, as a counter-example to (b) with compactness omitted, the open parallelotope

$$\left\{x \in \mathbf{R}^n \,\middle|\, x = \sum_{i=1}^{n} \lambda_i e_i, \quad -1 < \lambda_i < 1\right\}$$

built on the base $e = (e_1, \ldots, e_n)$, and the lattice having e as a $\mathbf{Z}$-base.

4.2. The canonical imbedding of a number field

Let K be a number field and let n be its degree. We have seen (Chapter II, § 4, Theorem 1) that there are n distinct isomorphisms $\sigma_i : K \to \mathbf{C}$. There are exactly n, because the minimal polynomial for a primitive element of K over $\mathbf{Q}$ (ibid., corollary of Theorem 1) has only n roots in $\mathbf{C}$. Let $\alpha : \mathbf{C} \to \mathbf{C}$ be complex conjugation. Then, for any $i = 1, \ldots, n$ $\alpha \circ \sigma_i = \sigma_j$, $1 \leq j \leq n$, and $\sigma_i = \sigma_j$ if and only if $\sigma_i(K) \subset \mathbf{R}$. Write r_1 for the number of indices such that $\sigma_i(K) \subset \mathbf{R}$. Then $n - r_1$ is an even number, so we may write

$$(1) \qquad r_1 + 2r_2 = n.$$

Let us renumber the σ_i's so that $\sigma_i(K) \subset \mathbf{R}$ for $1 \leq i \leq r_1$ and so that $\sigma_{j+r_2}(x) = \overline{\sigma_j(x)}$ for $r_1 + 1 \leq j \leq r_1 + r_2$. Then the first $r_1 + r_2$ isomorphisms determine the last r_2. For $x \in K$ we define

$$(2) \qquad \sigma(x) = (\sigma_1(x), \ldots, \sigma_{r_1+r_2}(x)) \in \mathbf{R}^{r_1} \times \mathbf{C}^{r_2}.$$

We call σ the *canonical imbedding* of K in $\mathbf{R}^{r_1} \times \mathbf{C}^{r_2}$; it is an injective ring homomorphism. We shall frequently identify $\mathbf{R}^{r_1} \times \mathbf{C}^{r_2}$ with $\mathbf{R}^n$ (cf. (1)). The notations σ, K, n, r_1, and r_2 will be in use for the rest of this chapter.

Proposition 1. If M *is a free* $\mathbf{Z}$*-submodule of* K *of rank n and if* $(x_i)_{1 \leq i \leq n}$ *is a* $\mathbf{Z}$*-base of* M, *then* $\sigma(M)$ *is a lattice in* $\mathbf{R}^n$, *whose volume is*

$$(3) \qquad v(\sigma(M)) = 2^{-r_2} \left| \det_{1 \leq i, j \leq n} (\sigma_i(x_j)) \right|.$$

Proof. For fixed i the coordinates of $\sigma(x_i)$ with respect to the canonical base of $\mathbf{R}^n$ are

$$(4) \quad \sigma_1(x_i), \ldots, \sigma_{r_1}(x_i),\ R(\sigma_{r_1+1}(x_i)),\ I(\sigma_{r_1+1}(x_i)),\ \ldots,\ R(\sigma_{r_1+r_2}(x_i)),\ I(\sigma_{r_1+r_2}(x_i)),$$

where R and I denote, respectively, the real and imaginary parts. Let us calculate the determinant D of the matrix whose ith column is given by (4). Making use of the formulas $R(z) = \frac{1}{2}(z + \bar{z})$ and $I(z) = (1/2i)(z - \bar{z})$ for $z \in \mathbf{C}$ and of the linearity of both R and I, we obtain $D = (2i)^{-r_2} \det(\sigma_j(x_i))$. Since the x_i's form a base for K over $\mathbf{Q}$, $\det(\sigma_j(x_i)) \neq 0$ (Chapter II, § 7, Proposition 3) and therefore $D \neq 0$. Thus the vectors $\sigma(x_i)$ are linearly independent in $\mathbf{R}^n$, so that the $\mathbf{Z}$-module which they generate (call it $\sigma(M)$) is a lattice in $\mathbf{R}^n$. The calculation of D given above shows that (3) does give its volume. Q.E.D.

Proposition 2. Let d be the absolute discriminant of K, *let* A *be the ring of integers in* K, *and let* $\mathfrak{a}$ *be a non-zero integral ideal of* A. *Then* $\sigma(\mathrm{A})$ *and* $\sigma(\mathfrak{a})$ *are lattices. Moreover,*

$$v(\sigma(\mathrm{A})) = 2^{-r_2}\,|d|^{1/2} \quad\text{and}\quad v(\sigma(\mathfrak{a})) = 2^{-r_2}\,|d|^{1/2}\,\mathrm{N}(\mathfrak{a}). \tag{5}$$

Proof. We know that A and $\mathfrak{a}$ are free **Z**-modules of rank n, so we may apply Proposition 1. On the other hand, if (x_i) is a **Z**-base for A, then $d = \det(\sigma_i(x_j))^2$ (Chapter II, § 7, Proposition 3). This proves the first formula in (5). The second formula follows from the first and the observation that $\sigma(\mathfrak{a})$ is a subgroup of $\sigma(\mathrm{A})$ of index $\mathrm{N}(\mathfrak{a})$ (Chapter III, § 5, Definition 1). A fundamental domain for $\sigma(\mathfrak{a})$ may obviously be constructed as the disjoint union of $\mathrm{N}(\mathfrak{a})$ copies of a fundamental domain for $\sigma(\mathrm{A})$. Q.E.D.

4.3. Finiteness of the ideal class group

Proposition 1. Let K *be a number field, n its degree,* r_1 *and* r_2 *the integers defined in the beginning of* § 2, *d the absolute discriminant of* K, *and* $\mathfrak{a}$ *a non-zero integral ideal of* K. *Then* $\mathfrak{a}$ *contains a non-zero element x such that*

$$|\mathrm{N}_{K/\mathbf{Q}}(x)| \le \left(\frac{4}{\pi}\right)^{r_2} \frac{n!}{n^n}\,|d|^{1/2}\,\mathrm{N}(\mathfrak{a}). \tag{1}$$

Proof. Let σ be the canonical imbedding of K into $\mathbf{R}^{r_1} \times \mathbf{C}^{r_2}$ (§ 2). Let t be a positive real number and let B_t be the set of all elements $(y_1, \ldots, y_{r_1}, z_1, \ldots, z_{r_2}) \in \mathbf{R}^{r_1} \times \mathbf{C}^{r_2}$ such that

$$\sum_{i=1}^{r_1} |y_i| + 2\sum_{j=1}^{r_2} |z_j| \le t. \tag{2}$$

Then B_t is a set which is compact and convex and symmetric relative to $0 \in \mathbf{R}^n$. By a calculation given in the appendix to this chapter

$$\mu(\mathrm{B}_t) = 2^{r_1}\left(\frac{\pi}{2}\right)^{r_2}\frac{t^n}{n!}. \tag{3}$$

Now choose t such that $\mu(\mathrm{B}_t) = 2^n v(\sigma(\mathfrak{a}))$, i.e. such that

$$2^{r_1}\left(\frac{\pi}{2}\right)^{r_2}\frac{t^n}{n!} = 2^{n-r_2}\,|d|^{1/2}\,\mathrm{N}(\mathfrak{a}) \quad (\text{§ 2, Proposition 2}),$$

or such that $t^n = 2^{n-r_1}\pi^{-r_2}n!.\,|d|^{1/2}\,\mathrm{N}(\mathfrak{a})$.

By the corollary to Theorem 2, § 1, there exists a non-zero element $x \in \mathfrak{a}$ such that $\sigma(x) \in \mathrm{B}_t$. Its norm has absolute value

$$|\mathrm{N}(x)| = \prod_{i=1}^{r_1} |\sigma_i(x)|\,.\prod_{j=r_1+1}^{r_1+r_2} |\sigma_j(x)|^2.$$

By the fact that the geometric mean never exceeds the arithmetic mean (cf. Hardy-Littlewood-Polya, "Inequalities", Cambridge University Press) we have

$$|\mathrm{N}(x)| \le \left[\frac{1}{n}\sum_{i=1}^{r_1} |\sigma_i(x)| + \frac{2}{n}\sum_{j=r_1+1}^{r_1+r_2} |\sigma_j(x)|\right]^n \le \frac{t^n}{n^n} \quad (\text{by (2)}).$$

Consequently,

$$|\mathrm{N}(x)| \leq \frac{1}{n^n} 2^{n-r_1} \pi^{-r_2} n! \, |d|^{1/2} \, \mathrm{N}(\mathfrak{a}),$$

which, combined with the relation $r_1 + 2r_2 = n$, yields (1). Q.E.D.

Corollary 1. With the same notations, every ideal class of K *(Chapter III, § 4) contains an integral ideal* $\mathfrak{b}$ *such that*

$$\mathrm{N}(\mathfrak{b}) \leq \left(\frac{4}{\pi}\right)^{r_2} \frac{n!}{n^n} |d|^{1/2}. \tag{4}$$

Proof. Let $\mathfrak{a}'$ be an ideal of the given class. By multiplying $\mathfrak{a}'$ by a principal ideal we may suppose that $\mathfrak{a} = \mathfrak{a}'^{-1}$ is an integral ideal. Take a non-zero element $x \in \mathfrak{a}$ for which (1) holds. Then $\mathfrak{b} = x\mathfrak{a}^{-1}$ is an integral ideal in the same class as $\mathfrak{a}'$, and $\mathrm{N}(\mathfrak{b})$ satisfies (4) by virtue of the multiplicativity of norms (Chapter III, § 5, Proposition 2). Q.E.D.

Corollary 2. Let K *be a number field, let* n *be its degree, and let* d *be its absolute discriminant. Then, for* $n \geq 2$,

$$|d| \geq \frac{\pi}{3}\left(\frac{3\pi}{4}\right)^{n-1}$$

and $n/(\log |d|)$ *is majorized by a constant independent of* K.

Since there is always a non-zero integral ideal $\mathfrak{b}$ in K and $\mathrm{N}(\mathfrak{b}) \geq 1$, we obtain, from (4), $|d|^{1/2} \geq (\pi/4)^{r_2} (n^n/n!)$. From $\pi/4 < 1$ and $2r_2 \leq n$ we conclude that $|d| \geq a_n$, where $a_n = (\pi/4)^n \, [n^{2n}/(n!)^2]$. We observe that

$$a_2 = \frac{\pi^2}{4} \quad \text{and} \quad \frac{a_{n+1}}{a_n} = \frac{\pi}{4}\left(1 + \frac{1}{n}\right)^{2n} = \frac{\pi}{4} \quad (1 + 2 + \text{positive terms})$$

(by use of the binomial formula), so $a_{n+1}/a_n \geq 3\pi/4$. Hence, for $n \geq 2$,

$$|d| \geq \frac{\pi^2}{4}\left(\frac{3\pi}{4}\right)^{n-2},$$

from which the inequality statement in the corollary is immediate. The uniform majoration of $n/(\log |d|)$ follows by taking logarithms.

Theorem 1 (Hermite-Minkowski). *For any number field* $\mathrm{K} \neq \mathbf{Q}$, *the absolute discriminant* d *of* K *is* $\neq \pm 1$.

Proof. Using Corollary 2 we see that $|d| \geq (\pi/3)(3\pi/4)^{n-1}$. Since $\pi/3 > 1$ and $3\pi/4 > 1$, we have $|d| > 1$. Q.E.D.

Theorem 2 (Dirichlet). *For any number field* K *the ideal class group is finite (Chapter III, § 4).*

Proof. By Corollary 1 of Proposition 1 it suffices to show that, for every positive integer q, the set of all integral ideals $\mathfrak{b}$ of K which have q as their norms is a finite set. For such an ideal $\mathfrak{b}$ we have $\mathrm{card}(\mathrm{A}/\mathfrak{b}) = q$ (Chapter III, § 5). It follows that $q \in \mathfrak{b}$, since for any group the order of an element divides the order of the group. Thus our ideals $\mathfrak{b}$ are among those which contain $\mathrm{A}q$, and there can be only finitely many such ideals (Chapter III, § 4, formula (8); or the finiteness of $\mathrm{A}/\mathrm{A}q$). Q.E.D.

Theorem 3 (Hermite). *In* $\mathbf{C}$ *there are only finitely many number fields with a given discriminant d.*

Proof. By Corollary 2 of Proposition 1 the degree of such a field is bounded. We may suppose n and the integers r_1 and r_2 are fixed. Let K be such a field.

In $\mathbf{R}^{r_1} \times \mathbf{C}^{r_2}$ consider the following set B:

(a) If $r_1 > 0$, B is the set of all elements $(y_1, \ldots, y_{r_1}, z_1, \ldots, z_{r_2}) \in \mathbf{R}^{r_1} \times \mathbf{C}^{r_2}$ such that

$$(5) \qquad |y_1| \leq 2^n \left(\frac{\pi}{2}\right)^{-r_2} |d|^{1/2};$$

$|y_i| \leq \frac{1}{2}$ for $i = 2, \ldots, r_1$; and $|z_j| \leq \frac{1}{2}$ for $j = 1, \ldots, r_2$.

(b) If $r_1 = 0$, B is the set of all elements $(z_1, \ldots, z_{r_2}) \in \mathbf{C}^{r_2}$ such that

$$(6) \quad |z_1 - \bar{z}_1| \leq 2^n \left(\frac{\pi}{2}\right)^{1-r_2} |d|^{1/2}, \quad |z_1 + \bar{z}_1| \leq \tfrac{1}{2}, \quad \text{and} \quad |z_j| \leq \tfrac{1}{2} \text{ for } j = 2, \ldots, r_2.$$

Clearly, B is a compact and convex set, which is symmetric about the origin in $\mathbf{R}^n$ and which has volume $2^{n-r_2}|d|^{1/2}$.[1] Writing σ for the canonical imbedding of K (§ 2), we obtain by means of Proposition 2 of § 2 and the corollary to Theorem 2 of § 1 an integer $x \neq 0$ of K for which $\sigma(x) \in$ B.

Let us show that x is a *primitive* element of K over $\mathbf{Q}$. In case (a), (5) shows that $|\sigma_i(x)| \leq \frac{1}{2}$ for $i \neq 1$. Since $|\mathrm{N}(x)| = \prod_{i=1}^{n} |\sigma_i(x)|$ is a positive integer (Chapter II, § 6, corollary of Proposition 2), we may conclude that $|\sigma_1(x)| \geq 1$, so $\sigma_1(x) \neq \sigma_i(x)$ for all $i \neq 1$. However, if x were not primitive, $\sigma_1(x)$ would coincide with $\sigma_i(x)$ for some $i \neq 1$ (Chapter II, § 6, Proposition 1). In case (b) we see similarly that $|\sigma_1(x)| = |\overline{\sigma_1(x)}| \geq 1$, so $\sigma_1(x) \neq \sigma_j(x)$ when σ_j is not σ_1 or $\overline{\sigma_1}$. Now (6) implies that the real part $\mathrm{R}(\sigma_1(x)) \leq \frac{1}{4}$. But this means that $\sigma_1(x)$ cannot be real, so $\sigma_1(x) \neq \overline{\sigma_1(x)}$. As in case (a) we conclude that x is primitive.

Formulas (5) and (6) imply that the conjugates $\sigma_i(x)$ of x are bounded. Therefore the elementary symmetric functions of the $\sigma_i(x)$'s are also bounded. In other words, the coefficients, as well as the degree, of the minimal polynomial of x are bounded. Since x is an integer, its minimal polynomial is a monic polynomial with coefficients in $\mathbf{Z}$ (Chapter II, § 6, the corollary of Proposition 2). The degree and the coefficients of the minimal polynomial of x being bounded, there are only finitely many possibilities for the minimal polynomial of x, consequently only finitely many possible values for $x \in \mathbf{C}$. As x generates K, there are only finitely many possibilities for K. Q.E.D.

4.4. The unit theorem

Let K be a number field and let A be the ring of integers in K. By abuse of language we use the expression "units of K" to refer to the units in the ring A. We remind the reader

1. One may calculate this volume by making use of the observation that B is a product of intervals, of discs, and of a rectangle in case (b).

that in any ring the units form a group under multiplication. We write A^* for the group of units in A.

The following result will be useful.

Proposition 1. Let K *be a number field and let* $x \in K$. *In order that* x *be a unit of* K *it is necessary and sufficient that* x *be an integer of* K *of norm* ± 1.

Proof. If x is a unit of K, then $N(x)$ and $N(x^{-1})$ belong to $\mathbf{Z}$. We have $N(x) \cdot N(x^{-1}) = N(x \cdot x^{-1}) = 1$, so $N(x) = \pm 1$. Conversely, let x be an integer of K with norm ± 1. Its characteristic equation has the form

$$x^n + a_{n-1}x^{n-1} + \cdots + a_1 x \pm 1 = 0 \quad \text{with } a_i \in \mathbf{Z} \text{ (Chapter II, § 6).}$$

Thus, $\pm(x^{n-1} + a_{n-1}x^{n-2} + \cdots + a_1) = x^{-1}$ and, since x^{-1} is an integer of K, x is a unit. Q.E.D.

Theorem 1 (Dirichlet). *Let* K *be a number field, n its degree, and let* r_1 *and* r_2 *be the integers defined in* § 2. *Set* $r = r_1 + r_2 - 1$. *The group* A^* *of units of* K *is isomorphic to* $\mathbf{Z}^r \times G$, *where* G *is a finite cyclic group comprised of the roots of unity contained in* K.

Proof. First we shall show that A^* is a commutative group of finite type. Then we shall calculate its rank.

Consider the canonical imbedding (§ 2) $x \mapsto (\sigma_1(x), \ldots, \sigma_{r_1+r_2}(x))$ of K into $\mathbf{R}^{r_1} \times \mathbf{C}^{r_2}$ and the mapping

$$x \mapsto L(x) = (\log|\sigma_1(x)|, \ldots, \log|\sigma_{r_1+r_2}(x)|) \tag{1}$$

of K^* to $\mathbf{R}^{r_1+r_2}$. (1) is a homomorphism (i.e. $L(xy) = L(x) + L(y)$), called the *logarithmic imbedding* of K^*. Let B be a compact subset of $\mathbf{R}^{r_1+r_2}$. Let us show that the set B′ of units $x \in A^*$ such that $L(x) \in B$ is a finite set. Indeed, since B is bounded, there exists a real number $\alpha > 1$ such that, for all $x \in B'$, $\alpha^{-1} \leq |\sigma_i(x)| \leq \alpha$ $(i = 1, \ldots, n)$. It follows that the elementary symmetric functions of the $\sigma_i(x)$'s are bounded in absolute value. Since they belong to $\mathbf{Z}$ (because $x \in A$), the set of possible values for the symmetric functions of the $\sigma_i(x)$'s is a finite set. Therefore, there are only finitely many possible characteristic polynomials for elements $x \in B'$ and consequently only finitely many possible values for x. Thus, B′ is a finite set.

The finiteness of B′ implies, immediately, the following statements:

(a) The kernel G of the restriction of L to A^* is a finite group. It consists, therefore, of roots of unity and is *cyclic* (Chapter I, § 6, Theorem 1). Clearly, every root of unity in K belongs to the kernel of L, for the roots of unity in K are integers and, if x is a root of unity in K, $|\sigma_i(x)|^q = |\sigma_i(x^q)| = |1| = 1$, so $|\sigma_i(x)| = 1$ for any i.

(b) The image $L(A^*)$ is a discrete subgroup of $\mathbf{R}^{r_1+r_2}$ (§ 1). Consequently, $L(A^*)$ is a free $\mathbf{Z}$-module of rank $s \leq r_1 + r_2$ (§ 1, Theorem 1). Since $L(A^*)$ is free, A^* is isomorphic to $G \times L(A^*) = G \times \mathbf{Z}^s$. It remains to show that the rank s of $L(A^*)$ equals $r_1 + r_2 - 1$.

The inequality $s \leq r_1 + r_2 - 1$ is easy. Indeed, for $x \in A^*$, the relation

$$\pm 1 = N(x) = \prod_{i=1}^{n} \sigma_i(x) = \prod_{i=1}^{r_1} \sigma_i(x) \prod_{j=r_1+1}^{r_1+r_2} \sigma_j(x)\overline{\sigma_j(x)} \quad \text{(Proposition 1)}$$

implies that the vector $L(x) = (y_1, \ldots, y_{r_1+r_2})$ lies in the hyperplane W defined by the equation

(2) $$\sum_{i=1}^{r_1} y_i + 2 \sum_{j=r_1+1}^{r_1+r_2} y_j = 0.$$

Since $L(A^*)$ is a discrete subgroup of W, $s \leq r_1 + r_2 - 1$.

Now we show that $L(A^*)$ contains $r = r_1 + r_2 - 1$ linearly independent vectors. This requires a more delicate argument. We are going to show that for any linear form $f \neq 0$ on W, there exists a unit u such that $f(L(u)) \neq 0$. As the projection of W on $\mathbf{R}^r$ is an isomorphism (by (2)), we may write, for any $y = (y_1, \ldots, y_{r+1}) \in W \subset \mathbf{R}^{r+1}$,

(3) $$f(y) = c_1 y_1 + \cdots + c_r y_r \quad \text{with } c_i \in \mathbf{R}.$$

Fix a real number α such that $\alpha \geq 2^n(1/2\pi)^{r_2}|d|^{1/2}$. For any set $\lambda = (\lambda_1, \ldots, \lambda_r)$ of r positive real numbers take $\lambda_{r+1} > 0$ such that $\prod_{i=1}^{r_1} \lambda_i \prod_{j=r_1+1}^{r_1+r_2} \lambda_j^2 = \alpha$. In $\mathbf{R}^{r_1} \times \mathbf{C}^{r_2}$ the set B of elements $(y_1, \ldots, y_{r_1}, z_1, \ldots, z_{r_2})$ ($y_i \in \mathbf{R}$ and $z_j \in \mathbf{C}$) such that $|y_i| \leq \lambda_i$ and $|z_j| \leq \lambda_{j+r_1}$ is compact, convex, symmetric about 0, and of volume $\prod_{i=1}^{r_1} 2\lambda_i \prod_{j=1+r_1}^{r_2+r_1} \pi\lambda_j^2 = 2^{r_1}\pi^{r_2}\alpha \geq 2^{n-r_2}|d|^{1/2}$. It follows from Proposition 2 of § 2 and the corollary to Theorem 2 of § 1 that there exists an integer x_λ of K such that $\sigma(x_\lambda) \in B$. This means that $|\sigma_i(x_\lambda)| \leq \lambda_i$ for $i = 1, \ldots, n$ (putting $\lambda_{j+r_2} = \lambda_j$ for $j = r_1 + 1, \ldots, r_1 + r_2$). Since x_λ is an integer,

$$1 \leq |N(x_\lambda)| = \prod_{i=1}^{n} |\sigma_i(x_\lambda)| \leq \prod_{i=1}^{r_1} \lambda_i \prod_{j=r_1+1}^{r_1+r_2} \lambda_j^2 = \alpha.$$

On the other hand, for any i,

$$|\sigma_i(x_\lambda)| = |N(x_\lambda)| \prod_{j \neq i} |\sigma_j(x_\lambda)|^{-1} \geq \prod_{j \neq i} \lambda_j^{-1} = \lambda_i\alpha^{-1}.$$

Now we have $\lambda_i\alpha^{-1} \leq |\sigma_i(x_\lambda)| \leq \lambda_i$ for any i, so that

(4) $$0 \leq \log \lambda_i - \log |\sigma_i(x_\lambda)| \leq \log \alpha.$$

Use of (3) entails

(5) $$|f(L(x_\lambda)) - \sum_{i=1}^{r} c_i \log \lambda_i| \leq (\sum_{i=1}^{r} |c_i|) \log \alpha.$$

Let β be a constant which is strictly larger than the right-hand side of (5) and, for every positive integer h, select r positive real numbers $\lambda_{i,h}$ $(i = 1, \ldots, r)$ satisfying $\sum_{i=1}^{r} c_i \log \lambda_{i,h} = 2\beta h$. Put $\lambda(h) = (\lambda_{1,h}, \ldots, \lambda_{r,h})$ and let x_h be the corresponding integer $x_{\lambda(h)}$. By (5) we have $|f(L(x_h)) - 2\beta h| < \beta$, so

(6) $$(2h - 1)\beta < f(L(x_h)) < (2h + 1)\beta.$$

It follows from (6) that the numbers $f(L(x_h))$ are all *distinct*. On the other hand, since $|N(x_h)| \leq \alpha$, there are only finitely many distinct ideals of the form Ax_h (cf. § 3, the proof of Theorem 2). Therefore, there exist two distinct indices h and k such that $Ax_h = Ax_k$ and, consequently, a unit $u \in A$ such that $x_k = ux_h$. We may conclude (since f is linear) that $f(L(u)) = f(L(x_k)) - f(L(x_h)) \neq 0$. Q.E.D.

Remark. Theorem 1 (called the "unit theorem") implies that there exist $r(=r_1+r_2-1)$ units (u_i) of K such that any unit u of K may be uniquely expressed in the form

$$u = zu_1^{n_1} \dots u_r^{n_r} \tag{7}$$

with $n_i \in \mathbf{Z}$ and z a root of unity. The set (u_i), $i = 1, \dots, r$, is called a *fundamental system of units of* K.

Example: cyclotomic fields. Let p be an odd prime number, let z be a primitive complex pth root of unity, and let K be the cyclotomic field $\mathbf{Q}[z]$ (cf. Chapter II, § 9). We know that $[\mathrm{K}:\mathbf{Q}] = p - 1$ (ibid., Theorem 1). Since no conjugate of z in $\mathbf{C}$ is real, $r_1 = 0$ and $2r_2 = p - 1$, so $r = (p-3)/2$.

4.5. Units in imaginary quadratic fields

Let K be an imaginary quadratic field (Chapter II, § 5). Then $r_1 = 0$, $2r_2 = 2$, $r_2 = 1$, and $r_1 + r_2 - 1 = 0$. Thus the only units in K are the roots of unity contained in K (§ 4, Theorem 1), a finite cyclic group. With a little calculation we shall prove this result directly and make it more precise.

Let $\mathrm{K} = \mathbf{Q}[\sqrt{-m}]$, where m is a square-free positive integer. Recall that the units of K are integers of norm ± 1 (§ 4, Proposition 1).

(1) If $m \equiv 1$ or $2 \pmod 4$, the ring of integers of K is $\mathbf{Z} + \mathbf{Z}\sqrt{-m}$ (Chapter II, § 5, Theorem 1). For $x = a + b\sqrt{-m}$ $(a, b \in \mathbf{Z})$ we have $\mathrm{N}(x) = a^2 + mb^2 \geq 0$. In order that x be a unit we must have $a^2 + mb^2 = 1$. If $m \geq 2$, this implies that $b = 0$ and $a = \pm 1$, so $x = \pm 1$. If $m = 1$, besides the solution $x = \pm 1$, there are the solutions $a = 0$, $b = \pm 1$, i.e. $x = \pm i$ (with $i^2 = -1$).

(2) If $m \equiv 3 \pmod 4$, the ring of integers of K is $\mathbf{Z} + \mathbf{Z}[(1+\sqrt{-m})/2]$ (Chapter II, § 5, Theorem 1). For $x = a + (b/2)(1+\sqrt{-m})$ $(a, b \in \mathbf{Z})$, we have $\mathrm{N}(x) = (a + b/2)^2 + mb^2/4$. In order that x be a unit we must have $(2a+b)^2 + mb^2 = 4$. If $m \geq 7$, this implies that $b = 0$, so $2a^2 = 4$, $a = \pm 1$, and $x = \pm 1$. If $m = 3$, the relations $b = \pm 1$ and $(2a \pm 1)^2 = \pm 1$ entail the additional solutions $x = \frac{1}{2}(\pm 1 \pm \sqrt{-3})$ (the signs $\pm$ being independent).

Summarizing, we have proved the following result:

Proposition 1. If K *is a quadratic imaginary field, the group* G *of units in* K *is comprised of the square roots of unity,* $+1$ *and* -1, *except in the following two cases:*

(1) *If* $\mathrm{K} = \mathbf{Q}[i]$ $(i^2 = -1)$, G *is comprised of the fourth roots of unity:* $i, -1, -i, 1$.

(2) *If* $\mathrm{K} = \mathbf{Q}[\sqrt{-3}]$, G *is comprised of the sixth roots of unity:* $[(1+\sqrt{-3})/2]^j$, $j = 0, 1, \dots, 5$.

4.6. Units in real quadratic fields

This section is going to be considerably more interesting than the preceding one. Let K be a real quadratic field. With the usual notations, we have $r_1 = 2$ and $r_2 = 0$, so $r = r_1 + r_2 - 1 = 1$. The unit theorem (§ 4, Theorem 1) implies that the group of units of K is isomorphic to the product of **Z** with the group of roots of unity contained in K. As K admits an imbedding into **R**, the only roots of unity are ± 1. Thus, assuming that K has been imbedded into **R**, we have:

Proposition 1. The positive units of a real quadratic field $K \subset \mathbf{R}$ *form a (multiplicative) group isomorphic to* **Z**.

This group contains one and only one generator larger than one; we call it *the fundamental unit of* K.

Let $K = \mathbf{Q}[\sqrt{d}]$, where $d \geq 2$ is a square-free integer, and let $x = a + b\sqrt{d}$ (a, $b \in \mathbf{Q}$) be a unit of K. The numbers x, x^{-1}, $-x$, and $-x^{-1}$ are units of K and, since $N(x) = (a + b\sqrt{d})(a - b\sqrt{d}) = \pm 1$ (§ 4, Proposition 1), these four numbers are $\pm a \pm b\sqrt{d}$. For $x \neq \pm 1$ only one of the four numbers x, x^{-1}, $-x$, $-x^{-1}$ is greater than one, and it is the largest of the four. Thus *the units greater than one of* K *are those of the form* $a + b\sqrt{d}$ *with* $a > 0$, $b > 0$.

(a) Suppose first that $d \equiv 2$ or $3 \pmod 4$. In this case the ring of integers of K is $\mathbf{Z} + \mathbf{Z}\sqrt{d}$ (Chapter II, § 5, Theorem 1). As the units of K are integers of norm ± 1 (§ 4, Proposition 1), the units greater than one of K are the numbers $a + b\sqrt{d}$ with a, $b \in \mathbf{Z}$ and $a > 0$, $b > 0$ such that

$$a^2 - db^2 = \pm 1. \tag{1}$$

We remark that the solutions "in natural numbers" (a, b) of equation (1) (called the "equation of Pell-Fermat") are obtained as follows: take the fundamental unit $a_1 + b_1\sqrt{d}$ of K and put

$$a_n + b_n\sqrt{d} = (a_1 + b_1\sqrt{d})^n, \quad n \geq 1. \tag{2}$$

The sequence (a_n, b_n) lists all the solutions of (1).

Remarks. (1) It follows from (2) that $b_{n+1} = a_1 b_n + b_1 a_n$. Since a_1, b_1, a_n and b_n are all positive, the sequence (b_n) is strictly increasing. Thus, in order to explicitly calculate the fundamental unit $a_1 + b_1\sqrt{d}$, it suffices to write down the sequence (db^2) for $b \in \mathbf{N}$, $b \geq 1$ and to stop at the first number db_1^2 of this sequence which differs by a square a_1^2 from ± 1. Then $a_1 + b_1\sqrt{d}$ is the fundamental unit of K. For instance, if $d = 7$, the sequence (db^2) is 7, 28, $63 = 64 - 1 = 8^2 - 1$, so, taking $b_1 = 3$ and $a_1 = 8$, we see that $8 + 3\sqrt{7}$ is the fundamental unit of $\mathbf{Q}[\sqrt{7}]$. We see similarly that the fundamental units of $\mathbf{Q}[\sqrt{2}]$, $\mathbf{Q}[\sqrt{3}]$, and $\mathbf{Q}[\sqrt{6}]$ are $1 + \sqrt{2}$, $2 + \sqrt{3}$, and $5 + 2\sqrt{6}$. Using the theory of continued fractions, one can find other, more rapid, procedures for calculating the fundamental unit.

(2) If the fundamental unit is of norm one, the sequence (a_n, b_n) gives solutions only for the equation $(1')\, a^2 - db^2 = 1$; in this case the equation $(1'')\, a^2 - db^2 = -1$ has no solution in natural numbers. If the fundamental unit has norm -1, the solutions

of (1′) comprise the sequence (a_{2n}, b_{2n}) and those of (1″) the sequence (a_{2n+1}, b_{2n+1}). The first case occurs when $d = 3, 6$, or 7, the second when $d = 2$ or 10 ($3 + \sqrt{10}$ is the fundamental unit in $\mathbf{Q}[\sqrt{10}]$).

(b) Assume now that $d \equiv 1 \pmod 4$. The integers of $K = \mathbf{Q}[\sqrt{d}]$ are the numbers $\frac{1}{2}(a + b\sqrt{d})$ with $a, b \in \mathbf{Z}$ of the same parity (Chapter II, § 5, Theorem 1). Consequently, if $\frac{1}{2}(a + b\sqrt{d})$ is a unit of K (§ 4, Proposition 2), we must have

$$a^2 - db^2 = \pm 4. \tag{3}$$

Conversely, if (a, b) is an integer solution of (3), then $\frac{1}{2}(a + b\sqrt{d})$ is an integer of K (its trace is a and its norm, by (3), is ± 1) and, hence, a unit of K. As in (a), writing $a_1 + b_1\sqrt{d}$ for the fundamental unit of K, we see that the solutions in pairs of natural numbers (a, b) of (3) comprise the values of the sequence (a_n, b_n) $(n \geq 1)$ defined by setting

$$a_n + b_n\sqrt{d} = 2^{1-n}(a_1 + b_1\sqrt{d})^n. \tag{4}$$

The calculation of $a_1 + b_1\sqrt{d}$ may be accomplished as in (a). For example, the fundamental units of $\mathbf{Q}[\sqrt{5}]$, $\mathbf{Q}[\sqrt{13}]$, and $\mathbf{Q}[\sqrt{17}]$ are $\frac{1}{2}(1 + \sqrt{5})$, $\frac{1}{2}(3 + \sqrt{13})$, and $4 + \sqrt{17}$; these three units all have norm -1. For the choice of the sign ± 1 in (3) we have results similar to those obtained in (a).

Remark. In the case $d \equiv 1 \pmod 4$ the solutions of the Pell-Fermat equation

$$a^2 - db^2 = \pm 1 \tag{5}$$

correspond to units $a + b\sqrt{d}(a, b > 0)$ which belong to the ring $B = \mathbf{Z}[\sqrt{d}]$. This ring B is a subring of the ring A of integers of K and the positive units of B form a subgroup G of the group of positive units of A. Let $u = \frac{1}{2}(a + b\sqrt{d})$ be the fundamental unit of K. If a and b are both even, then $u \in B$, so that G consists of the powers of u (this is the case, for instance, when $d = 17$). If a and b are both odd, then $u^3 \in B$. (To see this note that $8u^3 = a(a^2 + 3b^2d) + b(3a^2 + b^2d)\sqrt{d}$. Since $a^2 - b^2d = \pm 4$, $a^2 + 3b^2d = 4(b^2d \pm 1)$, which is a multiple of 8, since b and d are odd. Similarly $3a^2 + b^2d = 4(a^2 \pm 1)$, which is again, because a is odd, a multiple of 8.) In this case G is comprised of the powers of u^3 ($u^2 \notin B$, otherwise $u = u^3/u^2 \in B$). This happens, for instance, when $d = 5$ (respectively, $d = 13$), in which case $u^3 = 2 + \sqrt{5}$ (respectively, $u^3 = 18 + 5\sqrt{13}$).

4.7. A generalization of the unit theorem

Proposition 1. Let A *be a ring which is a* $\mathbf{Z}$*-module of finite type. Then the multiplicative group* A^* *consisting of the units of* A *is a (commutative) group of finite type.*

Remarks. For a commutative group G, "of finite type" means "of finite type with respect to the $\mathbf{Z}$-module structure of G". A subgroup of a commutative group of finite type is of finite type (Chapter III, § 1, Corollary 2 of Theorem 1). Let us note that A is a Noetherian ring, for the ideals of A are **Z**-submodules of A.

Proof of the proposition. First we treat the case when A is an integral domain. If its field of fractions K is of characteristic zero, K is a finite-dimensional vector space over **Q**, so K is a number field. On the other hand, A is integral over **Z** (since it is a **Z**-module of finite type, cf. Chapter II, § 1, Theorem 1), and, therefore, A is a subring of B, the ring of integers of K. Thus $A^* \subset B^*$ and, since B^* is of finite type by the unit theorem (§ 4, Theorem 1), so is A^*. If K is of characteristic $p \neq 0$, K is a finite extension of $\mathbf{F}_p$, so K is a finite field and A^* is a finite group.

Now let us consider the case in which A is *reduced* (by definition, this means that 0 is the only nilpotent element in A). We shall need the following lemma.

Lemma. In a reduced Noetherian ring A, *the ideal* (0) *is expressible as the intersection of finitely many prime ideals.*

Proof. We know that, in a Noetherian ring, any ideal contains a product of prime ideals (Chapter III, § 3, Lemma 3). (0) is the smallest ideal, so (0) *is* a product of prime ideals: $(0) = \mathfrak{p}_1^{n_1} \ldots \mathfrak{p}_q^{n_q}$. Let $x \in \mathfrak{p}_1 \cap \cdots \cap \mathfrak{p}_q$. Then $x^{n_1+\cdots+n_q} \in \mathfrak{p}_1^{n_1} \ldots \mathfrak{p}_q^{n_q} = (0)$, so $x^{n_1+\cdots+n_q} = 0$. Since A is reduced, this means that $x = 0$. Therefore, $\mathfrak{p}_1 \cap \cdots \cap \mathfrak{p}_q = (0)$. Q.E.D.

Now, returning to the proof of Proposition 1, we let $(0) = \mathfrak{p}_1 \cap \cdots \cap \mathfrak{p}_q$, the $\mathfrak{p}_i$'s being prime ideals. It follows that the canonical homomorphism $\varphi : A \to \prod_{i=1}^{q} A/\mathfrak{p}_i$ is injective. An element of a ring product is invertible if and only if all of its components are invertible, so that $(\prod_i A/\mathfrak{p}_i)^* = \prod_i (A/\mathfrak{p}_i)^*$. By the integral domain case, each $(A/\mathfrak{p}_i)^*$ is of finite type; therefore, $\prod (A/\mathfrak{p}_i)^*$ is of finite type and so is any subgroup, e.g. $\varphi(A^*)$ (recall that **Z** is a Noetherian ring). Since φ is injective, A^* is of finite type.

Let us finally consider the general case. Observe that the set $\mathfrak{n}$ of nilpotent elements of A is an ideal, since $x^p = 0$, $y^q = 0$, and $a \in A$ imply that $(x + y)^{p+q-1} = 0$ and $(ax)^p = 0$. On the other hand there exists an integer s such that $\mathfrak{n}^s = (0)$. (To see this, let $(x_1, \ldots, x_r)$ be a finite set of generators of the ideal $\mathfrak{n}$ in the Noetherian ring A. Assume $x_i^{q_i} = 0$, $i = 1, \ldots, r$. Then, with $s = q_1 + \cdots + q_r$, it is clear that $\mathfrak{n}^s = 0$.) We proceed by induction on s. The case $s = 1$ is the reduced case—already treated. Suppose $s > 1$ and write φ for the canonical homomorphism $\varphi : A \to A/\mathfrak{n}^{s-1}$. Then $\varphi(A^*) \subset (A/\mathfrak{n}^{s-1})^*$, so $\varphi(A^*)$ is of finite type. The kernel of φ restricted to A^* belongs to $1 + \mathfrak{n}^{s-1}$; in fact, $\ker(\varphi) = 1 + \mathfrak{n}^{s-1}$, because for $s > 1$, $(\mathfrak{n}^{s-1})^2 \subset \mathfrak{n}^s = (0)$, which implies that any element $1 + x \in 1 + \mathfrak{n}^{s-1}$ has $1 - x$ as its inverse: $(1 + x)(1 - x) = 1 - x^2 = 1$. To prove that $1 + \mathfrak{n}^{s-1}$ is of finite type we need only observe that the mapping $x \mapsto 1 + x$ of the additive group $\mathfrak{n}^{s-1}$ to $1 + \mathfrak{n}^{s-1}$ is an isomorphism ($\mathfrak{n}^{s-1}$ is of finite type because it is a submodule of A). But this is obvious from the relation $(1 + x)(1 + y) = 1 + x + y$ for $x, y \in \mathfrak{n}^{s-1}$. That A^* is of finite type now follows from Chapter III, § 1, Proposition 1. Q.E.D.

Using arguments borrowed from algebraic geometry, one can show that, for any *reduced* ring B of the form $B = \mathbf{Z}[x_1, \ldots, x_n]$ (i.e. finitely generated over **Z** as a ring), the group B^* of units in B is of finite type ([6]).

APPENDIX

The calculation of a volume

Proposition. Let r_1 *and* $r_2 \in \mathbf{N}$, $n = r_1 + 2r_2$, $t \in \mathbf{R}$, *and let* B_t *be the set of all elements* $(y_1, \ldots, y_{r_1}, z_1, \ldots, z_{r_2}) \in \mathbf{R}^{r_1} \times \mathbf{C}^{r_2}$ *such that*

$$\sum_{i=1}^{r_1} |y_i| + 2 \sum_{j=1}^{r_2} |z_j| \leq t. \tag{1}$$

Let μ *denote the Lebesgue measure in* $\mathbf{R}^n$. *Then,*

$$\mu(B_t) = 2^{r_1} \left(\frac{\pi}{2}\right)^{r_2} \frac{t^n}{n!} \quad \text{for any } t \geq 0. \tag{2}$$

Proof. We set $\mu(B_t) = V(r_1, r_2, t)$ and argue by double induction on r_1 and r_2. Clearly $V(1, 0, t) = 2t$ (the segment $[-t, +t]$) and $V(0, 1, t) = \pi t^2/4$ (the disc of radius $t/2$). These results verify (2) in the special cases considered.

Now assume (2) gives $V(r_1, r_2, t)$. First we compute $V(r_1 + 1, r_2, t)$. The set $B_t \subset \mathbf{R} \times \mathbf{R}^{r_1} \times \mathbf{C}^{r_2}$, which corresponds to $r_1 + 1$ and r_2, is defined by the relation

$$|y| + \sum_{i=1}^{r_1} |y_i| + 2 \sum_{j=1}^{r_2} |z_j| \leq t \quad \text{with } y \in \mathbf{R}.$$

Integrating "in strips" and observing that for $|y| > t$, $B_t = \phi$, we see that

$$V(r_1 + 1, r_2, t) = \int_{-t}^{+t} V(r_1, r_2, t - |y|)\, dy.$$

Use of the induction hypothesis gives

$$V(r_1 + 1, r_2, t) = 2 \int_0^t 2^{r_1} \left(\frac{\pi}{2}\right)^{r_2} \frac{(t - y)^n}{n!}\, dy = 2^{r_1+1} \left(\frac{\pi}{2}\right)^{r_2} \frac{t^{n+1}}{(n + 1)!},$$

which agrees with (2).

It remains to compute $V(r_1, r_2 + 1, t)$. We keep the induction hypothesis that $V(r_1, r_2, t)$ satisfies (2). The set $B_t \subset \mathbf{R}^{r_1} \times \mathbf{C}^{r_2} \times \mathbf{C}$ which corresponds to r_1 and $r_2 + 1$ is defined by the relation

$$\sum_{i=1}^{r_1} |y_i| + 2 \sum_{j=1}^{r_2} |z_j| + 2\,|z| \leq t \quad \text{with } z \in \mathbf{C}.$$

Again integrating in strips, we obtain

$$V(r_1, r_2 + 1, t) = \int_{|z| \leq t/2} V(r_1, r_2, t - 2\,|z|)\, d\mu(z),$$

where $d\mu(z)$ denotes Lebesgue measure on $\mathbf{C}$. Putting $z = \rho e^{i\theta}$ $(\rho \in \mathbf{R}_+, 0 \leq \theta \leq 2\pi)$, we have $d\mu(z) = \rho\, d\rho\, d\theta$. Use of the induction hypothesis gives

$$V(r_1, r_2 + 1, t) = \int_0^{t/2} \int_0^{2\pi} 2^{r_1} \left(\frac{\pi}{2}\right)^{r_2} \frac{(t - 2\rho)^n}{n!}\, \rho\, d\rho\, d\theta$$

$$= 2^{r_1} \left(\frac{\pi}{2}\right)^{r_2} \frac{2\pi}{n!} \int_0^{t/2} (t - 2\rho)^n \rho\, d\rho.$$

Calculating $\int_0^{t/2} (t - 2\rho)^n \rho \, d\rho$ by putting $2\rho = x$ and integrating by parts, we find that this integral has the value $t^{n+2}/[4(n + 1)(n + 2)]$. Thus,

$$V(r_1, r_2 + 1, t) = 2^{r_1} \left(\frac{\pi}{2}\right)^{r_2+1} \frac{t^{n+2}}{(n + 2)!},$$

which agrees with (2) since $r_1 + 2(r_2 + 1) = n + 2$.

Chapter v

The splitting of prime ideals in an extension field

Let K be a number field, A the ring of integers of K, L an extension of finite degree of K, and B the integral closure of A in L (i.e. the ring of integers of L). The ideal $B\mathfrak{p}$, generated in B by a non-zero prime ideal $\mathfrak{p}$ of A, is not in general prime. It splits into a product of prime ideals (Chapter III, § 4, Theorem 3), i.e. $B\mathfrak{p} = \prod_i \mathfrak{P}_i^{e_i}$. In this chapter we propose to study this splitting. The case in which B is a free A-module (for example, when A is a principal ideal ring; cf. Chapter II, § 7, the corollary of Theorem 1) is particularly easy. In § 1 we shall show how the general case may be reduced to this easier case.

5.1. Preliminaries concerning rings of fractions

Definition 1. Let A *be an integral domain, let* K *be its field of fractions, and let* S *be a subset of* A − (0) *which is stable under multiplication and contains* 1. *We call the set of all elements of* K *which may be written in the form* a/s *with* $a \in A$ *and* $s \in S$ *the ring of fractions of* A *with respect to* S. *We denote it* $S^{-1}A$.

Clearly $S^{-1}A$ is a commutative ring (since $a/s + a'/s' = (s'a + sa')/ss'$ and $(a/s)\cdot(a'/s') = aa'/ss'$) which contains A (since $1 \in S$). If $S = A - (0)$, then $S^{-1}A = K$. If S contains 1 alone, or if it contains only units of A, then $S^{-1}A = A$.

Proposition 1. Let A *be an integral domain and let* S *be a multiplicatively stable subset of* A − (0) *which contains* 1. *Set* $A' = S^{-1}A$.

(1) *For any ideal* $\mathfrak{b}'$ *of* A′, *it is true that* $(\mathfrak{b}' \cap A)A' = \mathfrak{b}'$, *so the mapping* $\mathfrak{b}' \mapsto \mathfrak{b}' \cap A$ *is an increasing (for the inclusion relation) injection of the set of ideals of* A′ *into the set of ideals of* A.
(2) *The mapping* $\mathfrak{p}' \mapsto \mathfrak{p}' \cap A$ *is an isomorphism of the partially ordered set (under inclusion) of prime ideals of* A′ *on the partially ordered set of prime ideals* $\mathfrak{p}$ *of* A *which satisfy* $\mathfrak{p} \cap S = \phi$. *The inverse mapping is* $\mathfrak{p} \mapsto \mathfrak{p}A'$.

Proof. 1. If $\mathfrak{b}'$ is an ideal of A′, then $\mathfrak{b}' \cap A \subset \mathfrak{b}'$ and $(\mathfrak{b}' \cap A)A' \subset \mathfrak{b}'$, since $\mathfrak{b}'$ is an ideal. To prove the reverse inclusion take $x \in \mathfrak{b}'$. Since $x = a/s$ ($a \in A$ and $s \in S$), $sx \in \mathfrak{b}'$ (since $A \subset A'$ and $\mathfrak{b}'$ is an ideal of A′). It follows that $a \in \mathfrak{b}'$ and $a \in \mathfrak{b}' \cap A$. Therefore,

$x = \frac{1}{s} a \in A'(\mathfrak{b}' \cap A)$, so $\mathfrak{b}' \subset A'(\mathfrak{b}' \cap A)$ and $\mathfrak{b}' = A'(\mathfrak{b}' \cap A)$. This formula assures the injectivity of the mapping $\varphi: \mathfrak{b}' \mapsto \mathfrak{b}' \cap A$, inasmuch as there is a mapping $\theta: \mathfrak{b} \mapsto A'\mathfrak{b}$ such that $\theta \circ \varphi =$ identity. It is clear that φ is an increasing mapping.

2. If $\mathfrak{p}'$ is a prime ideal of A′, then $\mathfrak{p} = \mathfrak{p}' \cap A$ is a prime ideal of A (Chapter III, § 3, Lemma 1). Furthermore, $\mathfrak{p} \cap S = \phi$, since, if $s \in \mathfrak{p} \cap S$, then $s \in \mathfrak{p}'$ and $1 = (1/s)s \in A'\mathfrak{p}' = \mathfrak{p}'$, an absurdity. Conversely, let $\mathfrak{p}$ be a prime ideal of A such that $\mathfrak{p} \cap S = \phi$. We are going to show that $\mathfrak{p}A'$ is a prime ideal of A′ and that $\mathfrak{p}A' \cap A = \mathfrak{p}$. Note first that $\mathfrak{p}A'$ is the set of all elements of A′ which are of the form p/s with $p \in \mathfrak{p}$ and $s \in S$; any element $x \in \mathfrak{p}A'$ may be written as

$$x = \sum_{i=1}^{n} \frac{a_i}{s_i} p_i (a_i \in A, s_i \in S, \text{and } p_i \in \mathfrak{p}),$$

so

$$x = \sum_{i=1}^{n} \frac{b_i}{s} p_i \left(s = s_1 \ldots s_n \text{ and } \frac{b_i}{s} = \frac{a_i}{s_i}\right).$$

Thus $x = p/s$ with $p = \Sigma b_i p_i \in \mathfrak{p}$. We may conclude that $1 \notin \mathfrak{p}A'$, since $\mathfrak{p} \cap S = \phi$ and since 1 cannot be written in the form p/s with $p \in \mathfrak{p}$ and $s \in S$. To show that the ideal $\mathfrak{p}A'$ is prime let $a/s \in A'$ and $b/t \in A'$ with $(a/s)\cdot(b/t) \in \mathfrak{p}A'$. Then $(a/s)\cdot(b/t) = (p/u)$ with $p \in \mathfrak{p}$ and $u \in S$, $abu = pst \in \mathfrak{p}$. Since $\mathfrak{p} \cap S = \phi$, $u \notin \mathfrak{p}$, so $ab \in \mathfrak{p}$ ($\mathfrak{p}$ is prime), which implies that either a or b belongs to $\mathfrak{p}$. Therefore, either a/s or b/s belongs to $\mathfrak{p}A'$, i.e. $\mathfrak{p}A'$ is prime. Let us show finally that $\mathfrak{p} = \mathfrak{p}A' \cap A$. Clearly, $\mathfrak{p} \subset \mathfrak{p}A' \cap A$. For the reverse inclusion take $x \in \mathfrak{p}A' \cap A$; then $x = p/s (p \in \mathfrak{p}, s \in S)$, since, by hypothesis, $x \in \mathfrak{p}A'$. Thus $sx = p \in \mathfrak{p}$; since $s \notin \mathfrak{p} (\mathfrak{p} \cap S = \phi)$ and since $\mathfrak{p}$ is prime, it follows that $x \in \mathfrak{p}$. Now we simply observe that the formulas $\mathfrak{p} = \mathfrak{p}A' \cap A$ and $\mathfrak{p}' = A'(\mathfrak{p}' \cap A)$ entail that the mappings $\varphi: \mathfrak{p}' \mapsto \mathfrak{p}' \cap A$ and $\theta: \mathfrak{p} \mapsto \mathfrak{p}A'$ (restricted to the prime ideals $\mathfrak{p}$ which do not intersect S) are inverse bijections, since $\theta \circ \varphi$ and $\varphi \circ \theta$ are both identity mappings. Q.E.D.

Corollary. If A *is a Noetherian integral domain, then every ring of fractions* $S^{-1}A$ *is Noetherian.*

Proof. By Proposition 1, (1) there is an injective mapping of the lattice of ideals in $S^{-1}A$ to a sublattice of the lattice of ideals in A. Therefore, the lattice of ideals in $S^{-1}A$ satisfies the maximal condition, so $S^{-1}A$ is a Noetherian ring.

Proposition 2. Let R *be an integral domain, let* A *be a subring of* R, *let* S *be a multiplicatively stable subset of* A *with* $1 \in S$ *and* $0 \notin S$, *and let* B *be the integral closure of* A *in* R. *Then the integral closure of* $S^{-1}A$ *in* $S^{-1}R$ *is* $S^{-1}B$.

Proof. Any element of $S^{-1}B$ may be written in the form b/s $(b \in B, s \in S)$. By dividing an equation of integral dependence for b on A, e.g.

$$b^n + a_{n-1}b^{n-1} + \cdots + a_0 = 0 \quad (a_i \in A),$$

by s^n we obtain an equation

$$\left(\frac{b}{s}\right)^n + \frac{a_{n-1}}{s}\left(\frac{b}{s}\right)^{n-1} + \cdots + \frac{a_0}{s^n} = 0,$$

which shows that b/s is integral over $S^{-1}A$. Conversely, if x/s ($x \in R, s \in S$), an element of $S^{-1}R$, is integral over $S^{-1}A$, then there is an equation of the form

$$\left(\frac{x}{s}\right)^n + \frac{a_{n-1}}{t_{n-1}}\left(\frac{x}{s}\right)^{n-1} + \cdots + \frac{a_0}{t_0} = 0 \quad (a_i \in A, t_i \in S).$$

Multiplying through by $(t_0 t_1 \ldots t_{n-1})^n$ shows that $xt_0 \ldots t_{n-1}/s$ is integral over A. Therefore, $xt_0 \ldots t_{n-1}/s \in B$ and $x/s = (1/t_0 \ldots t_{n-1})(xt_0 \ldots t_{n-1}/s)$ belongs to $S^{-1}B$. Q.E.D.

Corollary. If A *is an integrally closed ring, then every ring of fractions* $S^{-1}A$ *is integrally closed.*

Proof. Take for R in Proposition 2 the field of fractions of A.

Proposition 3. If A *is a Dedekind ring, every ring of fractions* $S^{-1}A$ *is a Dedekind ring.*

Proof. $S^{-1}A$ is Noetherian (the corollary of Proposition 1) and integrally closed (the corollary of Proposition 2). Furthermore, since one "loses" some prime ideals in passing from A to $S^{-1}A$ (Proposition 1, (2)), every non-zero prime ideal of $S^{-1}A$ is maximal.

Proposition 4. Let A *be a Dedekind ring, let* $\mathfrak{p}$ *be a non-zero prime ideal of* A, *and let* $S = A - \mathfrak{p}$. *Then* $S^{-1}A$ *is a principal ideal ring. More precisely, there exists a prime* $p \in S^{-1}A$ *such that the only non-zero ideals of* $S^{-1}A$ *are of the form* (p^n), $n \geq 0$.

Proof. Since $\mathfrak{p}$ is the only non-zero prime ideal of A disjoint from S, the only non-zero prime ideal of $S^{-1}A$ is $\mathfrak{P} = \mathfrak{p}S^{-1}A$ (Proposition 1, 2)). Since $S^{-1}A$ is a Dedekind ring (Proposition 3), its only non-zero ideals are of the form $\mathfrak{P}^n$ ($n \geq 0$). Take an element $p \in \mathfrak{P} - \mathfrak{P}^2$. The ideal (p) is contained in $\mathfrak{P}$ but not in $\mathfrak{P}^2$. Therefore, $(p) = \mathfrak{P}$ and $(p^n) = \mathfrak{P}^n$ for all $n \geq 0$. Thus $S^{-1}A$ is principal and all its ideals are of the form (p^n), $n \geq 0$. Q.E.D.

Proposition 5. Let A *be an integral domain,* S *a multiplicatively stable subset of* $A - (0)$ *containing 1, and let* $\mathfrak{m}$ *be a maximal ideal of* A *which is disjoint from* S. *Then* $S^{-1}A/\mathfrak{m}S^{-1}A \cong A/\mathfrak{m}$.

Proof. The composition of ring homomorphisms $A \to S^{-1}A \to S^{-1}A/\mathfrak{m}S^{-1}A$ has kernel $\mathfrak{m}S^{-1}A \cap A = \mathfrak{m}$, so there is an injection $\varphi: A/\mathfrak{m} \to S^{-1}A/\mathfrak{m}S^{-1}A$. We must show that φ is surjective. Take $x = a/s \in S^{-1}A$ ($a \in A, s \in S$) and let $\bar{x}$ denote its residue class in $S^{-1}A/\mathfrak{m}S^{-1}A$. Since $s \notin \mathfrak{m}$ (by hypothesis, $\mathfrak{m} \cap S = \phi$) and since $\mathfrak{m}$ is maximal, there exists $b \in A$ such that $bs \equiv 1 \pmod{\mathfrak{m}}$. Thus

$$\frac{a}{s} - ab = \frac{a}{s}(1 - bs) \in \mathfrak{m}S^{-1}A,$$

so $\varphi(ab) = \bar{x}$. Q.E.D.

5.2. The splitting of a prime ideal in an extension

In this section A *denotes a Dedekind ring of characteristic zero,* K *its field of fractions,* L *a finite extension of* K *of degree n, and* B *the integral closure of* A *in* L. *We remind the reader that* B *is also a Dedekind ring (Chapter III, § 4, Theorem 1).*

Let $\mathfrak{p}$ be a non-zero prime ideal of A. Then B$\mathfrak{p}$ is an ideal of B and it has an expression of the form

$$\mathrm{B}\mathfrak{p} = \prod_{i=1}^{q} \mathfrak{P}_i^{e_i}, \tag{1}$$

where the $\mathfrak{P}_i$'s are distinct prime ideals of B, the e_i's are positive integers, and the product sign denotes multiplication of ideals.

Proposition 1. The $\mathfrak{P}_i$*'s are precisely those prime ideals* $\mathfrak{Q}$ *of* B *such that* $\mathfrak{Q} \cap \mathrm{A} = \mathfrak{p}$.

Proof. For a prime ideal $\mathfrak{Q}$ of B the relation $\mathfrak{Q} \cap \mathrm{A} = \mathfrak{p}$ is equivalent to the relation $\mathfrak{Q} \supset \mathrm{B}\mathfrak{p}$ ($\Rightarrow$ is clear; $\Leftarrow$ follows from the fact that $\mathfrak{Q} \cap \mathrm{A}$ is a prime ideal of A and $\mathfrak{p}$ is maximal). Clearly, $\mathrm{B}\mathfrak{p} = \prod_{i=1}^{q} \mathfrak{P}_i^{e_i}$ implies $\mathrm{B}\mathfrak{p} \subset \mathfrak{P}_i$ for each $i = 1, \ldots, q$, so $\mathfrak{P}_i$ appears in the product expression for B$\mathfrak{p}$ if and only if $\mathfrak{P}_i \cap \mathrm{A} = \mathfrak{p}$. Q.E.D.

It is now clear that $\mathrm{A}/\mathfrak{p}$ may be identified with a subring of $\mathrm{B}/\mathfrak{P}_i$ for any $i = 1, \ldots, q$. Both rings are fields. Since B is an A-module of finite type (Chapter III, § 4, Theorem 1), $\mathrm{B}/\mathfrak{P}_i$ is a finite dimensional vector space over $\mathrm{A}/\mathfrak{p}$. We shall write f_i for the dimension of $\mathrm{B}/\mathfrak{P}_i$ over $\mathrm{A}/\mathfrak{p}$ and call f_i the *residual degree* of $\mathfrak{P}_i$ over A. The exponent e_i in (1) is called the *ramification index* of $\mathfrak{P}_i$ over A. Let us remark finally that $\mathrm{B}\mathfrak{p} \cap \mathrm{A} = \mathfrak{p}$ ($\supset$ clear; $\subset$ follows from the fact that, for each $i = 1, \ldots, q$, $\mathfrak{P}_i \cap \mathrm{A} = \mathfrak{p}$), so $\mathrm{B}/\mathrm{B}\mathfrak{p}$ is a vector space over $\mathrm{A}/\mathfrak{p}$, also of finite dimension.

Theorem 1. With the preceding notations

$$\sum_{i=1}^{q} e_i f_i = [\mathrm{B}/\mathrm{B}\mathfrak{p} : \mathrm{A}/\mathfrak{p}] = n. \tag{2}$$

Proof. The first equality is easy. Consider the sequence of ideals

$$\mathrm{B} \supset \mathfrak{P}_1 \supset \mathfrak{P}_1^2 \supset \cdots \supset \mathfrak{P}_1^{e_1} \supset \mathfrak{P}_1^{e_1}\mathfrak{P}_2 \supset \cdots \supset \mathfrak{P}_1^{e_1}\mathfrak{P}_2^{e_2} \supset \cdots \supset \mathfrak{P}_1^{e_1} \cdots \mathfrak{P}_q^{e_q} = \mathrm{B}\mathfrak{p}.$$

Two consecutive elements of this sequence are of the form $\mathfrak{B}$ and $\mathfrak{B}\mathfrak{P}_i$. Since there is no ideal strictly between $\mathfrak{B}$ and $\mathfrak{B}\mathfrak{P}_i$, $\mathfrak{B}/\mathfrak{B}\mathfrak{P}_i$ is a vector space of dimension one over $\mathrm{B}/\mathfrak{P}_i$ (cf. the proof of Proposition, § 5, Chapter III). Thus it is a vector space of dimension f_i over $\mathrm{A}/\mathfrak{p}$. For a given i there are exactly e_i consecutive elements of the above sequence with associated quotient space of the form $\mathfrak{B}/\mathfrak{B}\mathfrak{P}_i$, i.e. of dimension f_i over $\mathrm{A}/\mathfrak{p}$. The total dimension $[\mathrm{B}/\mathrm{B}\mathfrak{p} : \mathrm{A}/\mathfrak{p}]$ equals the sum of the dimensions of the quotients, so it is $\sum_{i=1}^{q} e_i f_i$.

The second equality is also easy in the case where B is a free A-module, in particular when A is *principal* (Chapter II, § 7, the corollary of Theorem 1). In this case a base $(x_1, \ldots, x_n)$ of B as an A-module gives, by reduction mod B$\mathfrak{p}$, a base for $\mathrm{B}/\mathrm{B}\mathfrak{p}$ over $\mathrm{A}/\mathfrak{p}$. We are going to reduce the general case to this case by considering the multiplicatively stable subset $\mathrm{S} = \mathrm{A} - \mathfrak{p}$ of A and the rings of fractions $\mathrm{A}' = \mathrm{S}^{-1}\mathrm{A}$ and $\mathrm{B}' = \mathrm{S}^{-1}\mathrm{B}$. We know that A′ is a principal ideal ring in which $\mathfrak{p}\mathrm{A}'$ is the unique maximal ideal (§ 1, Proposition 4), and that B′ is the integral closure of A′ in L (§ 1, Proposition 2). By the special case when A is principal, $[\mathrm{B}'/\mathfrak{p}\mathrm{B}' : \mathrm{A}'/\mathfrak{p}\mathrm{A}'] = n$. Now consider the factorization of the ideal $\mathfrak{p}B'$ in the Dedekind ring B': from the fact that $\mathfrak{p}B = \prod_{i=1}^{q} \mathfrak{P}_i^{e_i}$ we conclude that $\mathfrak{p}B' = \prod_{i=1}^{q} (B'\mathfrak{P}_i)^{e_i}$. Since $\mathfrak{P}_i \cap A = \mathfrak{p}$ (Proposition 1), $\mathfrak{P}_i \cap S = \phi$ and $B'\mathfrak{P}_i$ is a non-zero prime ideal of B' (§ 1, Proposition 1, (2)). From the first part of our proof we now obtain

$$[B'/\mathfrak{p}B' : A'/\mathfrak{p}A'] = \sum_{i=1}^{q} e_i[B'/B'\mathfrak{P}_i : A'/\mathfrak{p}A'].$$

However,

$$A'/\mathfrak{p}A' \cong A/\mathfrak{p} \quad \text{and} \quad B'/B'\mathfrak{P}_i \cong B/\mathfrak{P}_i \quad (\S\, 1, \text{Proposition } 5).$$

Therefore,

$$n = [B'/\mathfrak{p}B' : A'/\mathfrak{p}A'] = \sum_{i=1}^{q} e_i f_i.$$

Q.E.D.

Proposition 2. With the same notations, the ring $B/B\mathfrak{p}$ *is isomorphic to the ring* $\prod_{i=1}^{q} B\mathfrak{P}_i^{e_i}$.

Proof. $\mathfrak{P}_i$ is the only maximal ideal of B which contains $\mathfrak{P}_i^{e_i}$, so $\mathfrak{P}_i^{e_i} + \mathfrak{P}_j^{e_j} = B$ for $i \neq j$. The proposition now follows from (1) and Lemma 1, § 3, Chapter I.

Example. Cyclotomic fields. Let p be a prime number and let z be a primitive p^rth root of unity in the complex numbers. In this case, all the complex p^rth roots of unity are of the form z^j $(j = 1, \ldots, p^r)$. The primitive roots of unity are those for which j is not a multiple of p. The number of primitive roots is $\varphi(p^r) = p^r - p^{r-1} = p^{r-1}(p-1)$ (cf. Chapter I, § 6). The primitive p^rth roots of unity are the roots of the cyclotomic polynomial

$$(3) \qquad F(X) = \frac{X^{p^r} - 1}{X^{p^{r-1}} - 1} = X^{p^{r-1}(p-1)} + X^{p^{r-1}(p-2)} + \cdots + X^{p^{r-1}} + 1.$$

We intend to give another proof that $[\mathbf{Q}[z] : \mathbf{Q}] = p^{r-1}(p-1)$ i.e. that F is irreducible (cf. Chapter II, § 9). For this purpose put $e = p^{r-1}(p-1)$ and let $z_1, \ldots, z_e$ be all the primitive p^rth roots of unity. Since the constant term of $F(X+1)$ is p, $\prod_{j=1}^{e} (z_j - 1) = \pm p$. Let B be the ring of integers of $\mathbf{Q}[z]$. Clearly, $z_j \in B$ and $z_j - 1 \in B(z_k - 1)$, for all j and k, since z_j is a power z_k^q of z_k and $z_k^q - 1 = (z_k - 1)(z_k^{q-1} + \cdots + z_k + 1)$. Thus all the ideals $B(z_k - 1)$ are the same. It follows that $Bp = B(z_1 - 1)^e$.

Now write $Bp = \prod_{i=1}^{q} \mathfrak{P}_i^{e_i}$, where the $\mathfrak{P}_i$'s are prime ideals of B. The e_i's must, clearly, all be multiples of e. But $e \geq [\mathbf{Q}[z] : \mathbf{Q}]$ (by (3)), so $e \geq \sum_{i=1}^{q} e_i f_i$ (Theorem 1). From these inequalities (they are really equalities) we may conclude that $q = 1$, $e = e_1$, $f_1 = 1$, and $[\mathbf{Q}[z] : \mathbf{Q}] = e$. In summary:

(a) $[\mathbf{Q}[z] : \mathbf{Q}] = e = p^{r-1}(p-1)$,
(b) $B(z_1 - 1)$ is a prime ideal of B of residual degree 1, and
(c) $Bp = B(z_1 - 1)^e$.

5.3. The discriminant and ramification

With the notations of § 2 (let $B\mathfrak{p} = \prod_{i=1}^{q} \mathfrak{P}_i^{e_i}$) a prime ideal $\mathfrak{p}$ of A is said to *ramify* in B (or in L) if any one of its ramification indices e_i is larger than one. In terms of the theory of the discriminant (Chapter II, § 7) we are going to characterize those prime ideals of A which ramify in B. In particular we shall show that only finitely many prime ideals of A ramify in B. First we need some lemmas concerning the discrimant.

Lemma 1. Let A *be a ring, let* $B_1, \ldots, B_q$ *be rings containing* A *which are free* A*-modules of finite type, and let* $B = \prod_{i=1}^{q} B_i$ *be the product ring. Then* $\mathfrak{D}_{B/A} = \prod_{i=1}^{q} \mathfrak{D}_{B_i/A}$ *(cf. Chapter II, § 7, Definition 2).*

Proof. We formulate our proof for the case $q = 2$. The general statement follows by induction on q. Let $(x_1, \ldots, x_m)$ and $(y_1, \ldots, y_n)$ be bases for B_1 and B_2 as modules over A. With the usual identifications of B_1 with $B_1 \times (0)$ and B_2 with $(0) \times B_2$, we may consider $(x_1, \ldots, x_m, y_1, \ldots, y_n)$ as a base for $B = B_1 \times B_2$ over A. By definition of the product ring structure $x_i y_j = 0$, from which it follows that $\mathrm{Tr}(x_i y_j) = 0$. As a consequence the determinant $D(x_1, \ldots, x_m, y_1, \ldots, y_n)$ is the determinant of the matrix

$$\left|\begin{array}{c|c} \mathrm{Tr}(x_i x_{i'}) & 0 \\ \hline 0 & \mathrm{Tr}(y_j y_{j'}) \end{array}\right| .$$

The value of this determinant is

$$\det(\mathrm{Tr}(x_i x_{i'})) \cdot \det(\mathrm{Tr}(y_j y_{j'})),$$

so

$$D(x_1, \ldots, x_m, y_1, \ldots, y_n) = D(x_1, \ldots, x_m)\, D(y_1, \ldots, y_n).$$

Q.E.D.

Lemma 2. Let B *be a ring,* A *a subring of* B*, and* $\mathfrak{a}$ *an ideal of* A*. Assume that* B *is a free module over* A *with the base* $(x_1, \ldots, x_n)$*. For* $x \in B$ *write* $\bar{x}$ *for the residue class of* x *in* $B/\mathfrak{a}B$*. Then* $(\bar{x}_1, \ldots, \bar{x}_n)$ *is a base of* $B/\mathfrak{a}B$ *over* $A/\mathfrak{a}$ *and*

$$D(\bar{x}_1, \ldots, \bar{x}_n) = \overline{D(x_1, \ldots, x_n)}. \tag{1}$$

Proof. Let $x \in B$. If the matrix for multiplication by x, with respect to the base $(x_1, \ldots, x_n)$ is (a_{ij}) $(a_{ij} \in A)$, then the matrix for multiplication by $\bar{x}$ with respect to the base $(\bar{x}_1, \ldots, \bar{x}_n)$ is $(\bar{a}_{ij})$. Thus, $\mathrm{Tr}(\bar{x}) = \overline{\mathrm{Tr}(x)}$. Taking $x = x_i x_j$, we obtain $\mathrm{Tr}(\bar{x}_i \bar{x}_j) = \overline{\mathrm{Tr}(x_i x_j)}$, and (1) follows by taking determinants. Q.E.D.

Lemma 3. Let K *be a field which is finite or of characteristic zero. Let* L *be a finite dimensional (commutative)* K*-algebra. In order that* L *be reduced it is necessary and sufficient that* $\mathfrak{D}_{L/K} \neq (0)$.

Proof. Suppose first that L is not reduced and let x be a non-zero nilpotent element of L. Let $(x_1, \ldots, x_n)$ be a base for L over K such that $x = x_1$. Then $x_1 x_j$ is nilpotent and multiplication by $x_1 x_j$ is a nilpotent endomorphism of the vector space L over K. Thus, all the characteristic values of this endomorphism are zero, so $\mathrm{Tr}(x_1 x_j) = 0$.

The matrix $(\mathrm{Tr}(x_i x_j))$ has a row comprised entirely of zeroes, which implies that $D(x_1, \ldots, x_n) = 0$, i.e. $\mathfrak{D}_{L/K} = (0)$.

Conversely, suppose that L is reduced. Then the ideal (0) of L is expressible as a finite intersection of prime ideals, $(0) = \bigcap_{i=1}^{q} \mathfrak{P}_i$ (Chapter IV, § 7, Lemma). Since $L/\mathfrak{P}_i$ is an integral domain and a finite dimensional algebra over K, it is a field (Chapter II, § 1, Proposition 3). It follows that $\mathfrak{P}_i$ is a maximal ideal of L, and consequently $\mathfrak{P}_i + \mathfrak{P}_j = L$ for $i \neq j$. Therefore, L is isomorphic to the ring product $\prod_{i=1}^{q} L/\mathfrak{P}_i$ (Chapter I, § 3, Lemma 1). By Lemma 1 $\mathfrak{D}_{L/K} = \prod_{i=1}^{q} \mathfrak{D}_{(L/\mathfrak{P}_i)/K}$. But $\mathfrak{D}_{(L/\mathfrak{P}_i)/K} \neq (0)$, since K is finite or of characteristic zero (Chapter II, § 7, Proposition 3). Therefore, $\mathfrak{D}_{L/K} \neq (0)$.

Q.E.D.

Definition 1. Let K *and* L *be number fields with* $K \subset L$. *Let* A *and* B *be the rings of integers of* K *and* L, *respectively. The discriminant (ideal) of* B *over* A *(or of* L *over* K*) is the ideal of* A *generated by the discriminants of bases of* L *over* K *which are contained in* B. *Notation:* $\mathfrak{D}_{B/A}$ *or* $\mathfrak{D}_{L/K}$.

Remark 1. If $(x_1, \ldots, x_n)$ is a base of L over K contained in B, then $\mathrm{Tr}_{L/K}(x_i x_j) \in A$ (Chapter II, § 6, the corollary of Proposition 2), so $D(x_1, \ldots, x_n) \in A$. Thus $\mathfrak{D}_{B/A}$ is an integral ideal of A. It is non-zero by Chapter II, § 7, Proposition 3.

Remark 2. When B is a free A-module (for example if A is principal) we have already defined the discriminant $\mathfrak{D}_{B/A}$ as the ideal generated by $D(e_1, \ldots, e_n)$ where $(e_1, \ldots, e_n)$ is an A-module base for B (Chapter II, § 7, Definition 2). Our old definition coincides with the one given above, since, given any base (x_i) of L over K contained in B, one sees that $x_i = \sum_j a_{ij} e_j$ with $a_{ij} \in A$. Therefore,

$$D(x_1, \ldots, x_n) = \det(a_{ij})^2 \, D(e_1, \ldots, e_n)$$

(Chapter II, § 7, Proposition 1).

Theorem 1. Let the notations be as in the definition. In order that a prime ideal $\mathfrak{p}$ *of* A *ramify in* B, *it is necessary and sufficient that it contain the discriminant* $\mathfrak{D}_{B/A}$. *There are only finitely many prime ideals of* A *which ramify in* B.

Proof. The second assertion follows from the first, since $\mathfrak{D}_{B/A} \neq (0)$. Let us prove the first. Since $B/\mathfrak{p}B \simeq \prod_{i=1}^{q} B/\mathfrak{P}_i^{e_i}$ (§ 2, Proposition 2), the assertion "$\mathfrak{p}$ ramifies" is equivalent to the statement "$B/\mathfrak{p}B$ is not reduced", i.e. equivalent to "$\mathfrak{D}_{(B/\mathfrak{p}B)/(A/\mathfrak{p})} = (0)$" (by Lemma 3 and the fact that $A/\mathfrak{p}$ is a finite field). Now put $S = A - \mathfrak{p}$, $A' = S^{-1}A$, $B' = S^{-1}B$, and $\mathfrak{p}' = \mathfrak{p}A'$. Then A' is a principal ideal ring (§ 1, Proposition 4), B' is a free A'-module, $A/\mathfrak{p} \simeq A'/\mathfrak{p}'$, and $B/\mathfrak{p}B \simeq B'/\mathfrak{p}'B'$ (§ 1, Proposition 5). Therefore, writing $(e_1, \ldots, e_n)$ for an A'-module base of B', we know that $\mathfrak{D}_{(B/\mathfrak{p}B)/(A/\mathfrak{p})} = (0)$ if and only if $D(e_1, \ldots, e_n) \in \mathfrak{p}'$ (Lemma 2). If $D(e_1, \ldots, e_n) \in \mathfrak{p}'$ and if $(x_1, \ldots, x_n)$

is a base for L over K contained in B, then $x_i = \Sigma a'_{ij} e_j$ with $a'_{ij} \in A'$ (for $B \subset B'$), so

$$D(x_1, \ldots, x_n) = \det(a'_{ij})^2 D(e_1, \ldots, e_n) \in \mathfrak{p}'.$$

Since $\mathfrak{p}' \cap A = \mathfrak{p}$ (§ 1, Proposition 1, (2)), we may conclude that $D(x_1, \ldots, x_n) \in \mathfrak{p}$ and $\mathfrak{D}_{B/A} \subset \mathfrak{p}$. Conversely, if $\mathfrak{D}_{B/A} \subset \mathfrak{p}$, then $D(e_1, \ldots, e_n) \in \mathfrak{p}'$, for one may write $e_i = y_i s^{-1}$ with $y_i \in B$ and $s \in S$, for $1 \leq i \leq n$. Consequently,

$$D(e_1, \ldots, e_n) = s^{-2n} D(y_1, \ldots, y_n) \in A'\mathfrak{D}_{B/A} \subset A'\mathfrak{p} = \mathfrak{p}'.$$

Q.E.D.

Example 1. Quadratic fields. Let $K = \mathbf{Q}$ and $L = \mathbf{Q}[\sqrt{d}]$, where d is a square-free integer (Chapter II, § 5).

(a) If $d \equiv 2$ or $3 \pmod 4$, then $(1, \sqrt{d})$ is a base for the ring of integers of L. As $\mathrm{Tr}(1) = 2$, $\mathrm{Tr}(\sqrt{d}) = 0$, and $\mathrm{Tr}(d) = 2d$, it follows that $D(1, \sqrt{d}) = 4d$. The prime numbers which ramify in L therefore include 2 and the prime divisors of d.

(b) If $d \equiv 1 \pmod 4$, $(1, (1 + \sqrt{d})/2)$ is a base for the ring of integers of L. We see that

$$\mathrm{Tr}(1) = 2, \quad \mathrm{Tr}\left(\frac{1 + \sqrt{d}}{2}\right) = 1,$$

and

$$\mathrm{Tr}\left(\left(\frac{1 + \sqrt{d}}{2}\right)^2\right) = \mathrm{Tr}\left(\frac{d + 1}{4} + \frac{1}{2}\sqrt{d}\right) = \frac{d + 1}{2};$$

thus

$$D\left(1, \frac{1 + \sqrt{d}}{2}\right) = 2 \cdot \frac{d + 1}{2} - 1 = d.$$

The only prime numbers which ramify in L are the divisors of d.

We remark that a quadratic field $\mathbf{Q}[\sqrt{d}]$ is uniquely determined by its discriminant. In fact,

$D \equiv 0 \pmod 4$ implies $d = \dfrac{D}{4}$ (we must have $d \equiv 2$ or $3 \pmod 4$),

$D \equiv 1 \pmod 4$ implies $d = D$,

and $D \equiv 2$ or $3 \pmod 4$ is impossible.

Example 2. Cyclotomic fields. Let p be an odd prime number, let z be a primitive complex pth root of unity, and let $L = \mathbf{Q}[z]$ be the corresponding cyclotomic field. We know that $(1, z, \ldots, z^{p-2})$ is a $\mathbf{Z}$-base for the ring of integers B of L (Chapter II, § 9, Theorem 2), and that the minimal polynomial $F(X)$ of z over $\mathbf{Q}$ satisfies the relation $(X - 1)F(X) = X^p - 1$ (ibid., Theorem 1). We are going to calculate the discriminant $\mathfrak{D}_{B/\mathbf{Z}}$ by making use of the formula

$$D(1, z, \ldots, z^{p-2}) = (-1)^{\frac{1}{2}(p-1)(p-2)} N(F'(z)) \quad \text{(Chapter II, § 7, formula (6))}.$$

By taking the derivative of both sides of the relation $(X - 1)F(X) = X^p - 1$, we obtain $(z-1)F'(z) = pz^{p-1}$ (since $F(z) = 0$). We know that $N(p) = p^{p-1}$, $N(z) = \pm 1$, and $N(z-1) = \pm p$ (Chapter II, § 9). Therefore,

$$D(1, z, \ldots, z^{p-2}) = \pm p^{p-2}. \tag{1}$$

It follows that p is the only prime number which ramifies in $\mathbf{Q}[z]$.

The following result is sometimes useful for determining the ring of integers of a number field.

Proposition 1. Let L *be a number field of degree n over* $\mathbf{Q}$ *and let* $(x_1, \ldots, x_n)$ *be a* $\mathbf{Q}$*-base for* L *contained in the ring* B *of integers of* L. *If the discriminant* $D(x_1, \ldots, x_n)$ *is square-free, then* $(x_1, \ldots, x_n)$ *is a* $\mathbf{Z}$*-base for* B.

Proof. If $(e_1, \ldots, e_n)$ is a $\mathbf{Z}$-base for B, then $x_i = \sum_{j=1}^{n} a_{ij}e_j$, with $a_{ij} \in \mathbf{Z}$, whence

$$D(x_1, \ldots, x_n) = \det(a_{ij})^2 D(e_1, \ldots, e_n).$$

Since $D(x_1, \ldots, x_n)$ is square-free, $\det(a_{ij})$ must be ± 1, so $(x_1, \ldots, x_n)$ is also a $\mathbf{Z}$-base for B. Q.E.D.

Remark. The cyclotomic fields (for $p \geq 5$), or quadratic fields, provide examples which show that the sufficient condition of Proposition 1 is not a necessary condition.

Example. The polynomial $X^3 - X - 1$ (respectively, $X^3 + X + 1$, $X^3 + 10X + 1$) is irreducible over $\mathbf{Q}$. Otherwise it would have a linear factor, i.e. a rational root $x \in \mathbf{Q}$. But, the polynomial is monic, so $x \in \mathbf{Q}$ implies $x \in \mathbf{Z}$. The constant term of the polynomial is 1, so $x = \pm 1$ (x has to divide the constant term). Checking that neither ± 1 satisfies the polynomial, we may conclude that the polynomial is irreducible.

Now let x be a complex root of the given polynomial. The field $L = \mathbf{Q}[x]$ is a cubic field (i.e. of degree 3). Thus, $(1, x, x^2)$ is a base for L over $\mathbf{Q}$, and x is, clearly, an integer of L. By formula (7) of § 7, Chapter II, $D(1, x, x^2) = 4 - 27 = -23$ (respectively -31, -4027), the negative of a prime number. Therefore, $(1, x, x^2)$ is a $\mathbf{Z}$-base for the ring of integers of L.

5.4. The splitting of a prime number in a quadratic field

Let $d \in \mathbf{Z}$ be square-free, let L be the quadratic field $\mathbf{Q}[\sqrt{d}]$, write B for the ring of integers of L, and let p be a prime number. We are going to study the factorization of the ideal pB into a product of prime ideals of B.

The formula $\sum_{i=1}^{q} e_i f_i = 2$ (§ 2, Theorem 1) entails $q \leq 2$ and the following three possibilities:

(a) $q = 2$, $e_1 = e_2 = 1$, $f_1 = f_2 = 1$;
in this case we say that p *splits* in L.

(b) $q = 1$, $e_1 = 1$, $f_1 = 2$;
in this case we say that p *remains prime* in L.

(c) $q = 1$, $e_1 = 2$, $f_1 = 1$;
this means that p *ramifies* in L.

Let us first consider the case in which p *is odd.* We know (Chapter II, § 5) that

$B = \mathbf{Z} + \mathbf{Z}\sqrt{d}$ or $B = \mathbf{Z} + \mathbf{Z}[(1 + \sqrt{d})/2]$, depending upon d. But, if we pass to the residue classes of B modulo Bp, we see in the second case that $a + b[(1 + \sqrt{d})/2]$ (with b odd) is congruent to $a + (b + p)[(1 + \sqrt{d})/2]$, which belongs to $\mathbf{Z} + \mathbf{Z}\sqrt{d}$. Thus, for any d, we have $B/Bp \cong (\mathbf{Z} + \mathbf{Z}\sqrt{d})/(p)$. Also we see that

$$\mathbf{Z} + \mathbf{Z}\sqrt{d} \cong \mathbf{Z}[X]/(X^2 - d),$$

so

$$B/Bp \cong \mathbf{Z}[X]/(p, X^2 - d) \cong (\mathbf{Z}[X]/(p))/(X^2 - \bar{d}) \cong \mathbf{F}_p[X]/(X^2 - \bar{d}),$$

where $\bar{d}$ denotes the residue class of d modulo p. Now the assertion that p splits (respectively, remains prime, ramifies) in B has the interpretation: B/Bp is the product of two fields (respectively, is a field, contains nilpotent elements) (cf. § 2, Proposition 2). In other words, the polynomial $X^2 - \bar{d} \in \mathbf{F}_p[X]$ is the product of two distinct linear polynomials (respectively, is irreducible, is a square). This happens if $\bar{d}$ is a non-zero square in $\mathbf{F}_p$ (respectively, is not a square in $\mathbf{F}_p$, is zero in $\mathbf{F}_p$). When $\bar{d}$ is a non-zero square in $\mathbf{F}_p$ (respectively, is not a square in $\mathbf{F}_p$), we say that d is a *quadratic residue* (respectively, *non-residue*) modulo p.

Let us now consider the case $p = 2$. If $d \equiv 2$ or $3 \pmod 4$, then $B = \mathbf{Z} + \mathbf{Z}\sqrt{d}$, so, as above, $B/2B \cong \mathbf{F}_2[X]/(X^2 - \bar{d})$. In this case $X^2 - \bar{d}$ equals X^2 or $X^2 + 1 = (X + 1)^2$, in either case a square. Thus 2 ramifies in B. If $d \equiv 1 \pmod 4$, $(1 + \sqrt{d})/2$ has $X^2 - X - (d - 1)/4$ as its minimal polynomial, so, as above, $B/2B \cong [\mathbf{F}_2[X]/(X^2 - X - \delta)]$, where δ is the residue class mod 2 of $(d - 1)/4$. For $d \equiv 1 \pmod 8$, $\delta = 0$ and $X^2 - X - \delta = X(X - 1)$, so that 2 splits. For $d \equiv 5 \pmod 8$, $\delta = 1$ and $X^2 - X - \delta = X^2 + X + 1$, which is irreducible in $\mathbf{F}_2[X]$, so 2 remains prime.

In summary, we have proved the following:

Proposition 1. Let $L = \mathbf{Q}[\sqrt{d}]$, *the quadratic field associated with the square-free integer d.*

(a) *The odd primes p for which d is a quadratic residue* mod p *split in* L. *So does* 2, *if* $d \equiv 1 \pmod 8$.
(b) *The odd primes p for which d is not a quadratic residue* mod p *remain prime in* L. *So does* 2, *if* $d \equiv 5 \pmod 8$.
(c) *The odd prime divisors of d ramify in* L. *So does* 2, *if* $d \equiv 2$ *or* $3 \pmod 4$.

Part (c) was proved earlier, as an example in § 3.

5.5. The quadratic reciprocity law

Given an *odd* prime p and an integer d relatively prime to p we introduced in § 4 the locution "d is a quadratic residue mod p" (respectively, "d is a non-residue mod p") as meaning that the residue class of d mod p is a square (respectively, not a square) in $\mathbf{F}_p^*$. Now we define the *Legendre symbol* as follows:

$$(1) \qquad \begin{cases} \left(\dfrac{d}{p}\right) = +1, & \text{if } d \text{ is a quadratic residue mod } p, \\[2mm] \left(\dfrac{d}{p}\right) = -1, & \text{if } d \text{ is a non-residue mod } p. \end{cases}$$

It is understood that $\left(\frac{d}{p}\right)$ is defined only for integers d which are relatively prime to p, i.e. $d \in \mathbf{Z}-p\mathbf{Z}$. The multiplicative group $\mathbf{F}_p^*$ being cyclic of even order $p-1$ (Chapter I, § 7, Theorem 1), the squares in $\mathbf{F}_p^*$ form a subgroup $(\mathbf{F}_p^*)^2$ of index 2, and $\mathbf{F}_p^*/(\mathbf{F}_p^*)^2$ is isomorphic to $\{+1, -1\}$. Clearly, the Legendre symbol stands for the composition of the following homomorphisms:

$$\mathbf{Z}-p\mathbf{Z} \to \mathbf{F}_p^* \to \mathbf{F}_p^*/(\mathbf{F}_p^*)^2 \cong \{+1, -1\}.$$

As a consequence there is the formula:

$$\left(\frac{ab}{p}\right) = \left(\frac{a}{p}\right)\left(\frac{b}{p}\right), \quad a, b \in \mathbf{Z}-p\mathbf{Z}. \tag{2}$$

Proposition 1 ("Euler's criterion"). *If p is an odd prime and if $a \in \mathbf{Z}-p\mathbf{Z}$, then*

$$\left(\frac{a}{p}\right) \equiv a^{(p-1)/2} \pmod{p}.$$

Proof. Write w for a primitive root mod p (Chapter I, § 7). Then $a \equiv w^j \pmod{p}$, with $0 \leq j \leq p-2$, since the residue class $\bar{w}$ of w generates $\mathbf{F}_p^*$. Clearly, a is a quadratic residue if and only if j is even. Therefore, $\left(\frac{a}{p}\right) = (-1)^j$. On the other hand, $\mathbf{F}_p^*$ contains only one element of order 2; this element can be written either as $\bar{w}^{(p-1)/2}$ or -1. In $\mathbf{Z}$, we have $-1 \equiv w^{(p-1)/2} \pmod{p}$. Thus,

$$\left(\frac{a}{p}\right) = (-1)^j \equiv w^{j(p-1)/2} \equiv a^{(p-1)/2} \pmod{p}.$$

Q.E.D.

Now we are going to prove a famous theorem, which provides us with a relation between the Legendre symbols for distinct odd primes.

Theorem 1 ("The quadratic reciprocity law of Legendre-Gauss"). *If p and q are distinct odd prime numbers, then*

$$\left(\frac{p}{q}\right)\left(\frac{q}{p}\right) = (-1)^{(p-1)(q-1)/4}$$

Proof. Let us consider, in a suitable extension of $\mathbf{F}_q$, a primitive pth root of unity w. Since $w^p = 1$, the expression w^x is well defined for $x \in \mathbf{F}_p$. We shall also consider the Legendre symbol $\left(\frac{x}{p}\right)$ as defined on $x \in \mathbf{F}_p^*$, for $\left(\frac{d}{p}\right)$ clearly depends only on the residue class of d mod p. For $a \in \mathbf{F}_p^*$ consider the "Gaussian sum"

$$\tau(a) = \sum_{x \in \mathbf{F}_p^*} \left(\frac{x}{p}\right) w^{ax}. \tag{3}$$

It is an element of an extension field of $\mathbf{F}_q$. Putting $ax = y$, we have

$$\tau(a) = \sum_{y \in \mathbf{F}_p^*} \left(\frac{ya^{-1}}{p}\right) w^y = \left(\frac{a^{-1}}{p}\right) \sum_{y \in \mathbf{F}_p^*} \left(\frac{y}{p}\right) w^y \quad \text{(by (2))},$$

so

$$\tau(a) = \left(\frac{a}{p}\right) \tau(1). \tag{4}$$

On the other hand, since we are working in a field of characteristic q and since $\left(\frac{x}{p}\right) \in \mathbf{F}_q$, we have $\tau(1)^q = \sum_{x \in \mathbf{F}_p^*} \left(\frac{x}{p}\right)^q w^{qx}$, so, identifying q with its residue class mod p, we obtain

$$\tau(1)^q = \tau(q). \tag{5}$$

Let us calculate $\tau(1)^2$. We have

$$\tau(1)^2 = \sum_{x, y \in \mathbf{F}_p^*} \left(\frac{x}{p}\right)\left(\frac{y}{p}\right) w^{x+y}.$$

Putting $y = tx$, we see that

$$\tau(1)^2 = \sum_{x, t \in \mathbf{F}_p^*} \left(\frac{x}{p}\right)^2 \left(\frac{t}{p}\right) w^{x(1+t)} = \sum_{x,t} \left(\frac{t}{p}\right) w^{x(1+t)} = \sum_{t \in \mathbf{F}_p^*} \left[\left(\frac{t}{p}\right) \sum_{x \in \mathbf{F}_p^*} w^{x(1+t)}\right].$$

If $w^{1+t} \neq 1$, then $\sum_{j=0}^{p-1} (w^{1+t})^j = 0$ by the formula for summing a geometric series. Since $(w^{1+t})^0 = 1$, we have

$$\sum_{x \in \mathbf{F}_p^*} w^{x(1+t)} = -1, \text{ if } w^{1+t} \neq 1.$$

If $w^{1+t} = 1$, then $\sum_{x \in \mathbf{F}_p^*} w^{x(1+t)} = p - 1$; and this happens if and only if $t = -1$, since w is a primitive pth root of unity. Therefore,

$$\tau(1)^2 = \left(\frac{-1}{p}\right)(p-1) - \sum_{\substack{t \in \mathbf{F}_p^* \\ t \neq -1}} \left(\frac{t}{p}\right).$$

As there are as many squares as non-squares in $\mathbf{F}_p^*$,

$$\sum_{t \in \mathbf{F}_p^*} \left(\frac{t}{p}\right) = 0, \quad \text{so} \quad \tau(1)^2 = \left(\frac{-1}{p}\right)(p-1) + \left(\frac{-1}{p}\right) = \left(\frac{-1}{p}\right)p.$$

Using Euler's criterion (Proposition 1), we obtain

$$\tau(1)^2 = (-1)^{(p-1)/2}\, p. \tag{6}$$

Finally, using (4) and (5), we see that

$$\tau(1)^q = \tau(q) = \left(\frac{q}{p}\right)\tau(1).$$

By (6), $\tau(1) \neq 0$, so we may simplify:

$$\tau(1)^{q-1} = \left(\frac{q}{p}\right).$$

Use of (6) again yields

$$\left(\frac{q}{p}\right) = (\tau(1)^2)^{(q-1)/2} = (-1)^{(p-1)(q-1)/4}\, p^{(q-1)/2}.$$

Since $p^{(q-1)/2} = \left(\frac{p}{q}\right)$ (Proposition 1) and since $\left(\frac{p}{q}\right) = \left(\frac{p}{q}\right)^{-1}$, we have proved the quadratic reciprocity law. Q.E.D.

Proposition 2 ("the complementary formulas"). *If p is an odd prime, then*

(a) $\left(\frac{-1}{p}\right) = (-1)^{(p-1)/2}$ and

(b) $\left(\frac{2}{p}\right) = (-1)^{(p^2-1)/8}$.

Proof. Relation (a) is a special case of Euler's criterion (Proposition 1). We need only prove (b). Note first that, since the squares of 1, 3, 5, and 7 mod 8 are 1, 1, 1, and 1, $p^2 \equiv 1 \pmod 8$, so formula (b) makes sense. Let us remark next that, in the group $H = \{1, 3, 5, 7\}$ of units in the ring $\mathbf{Z}/8\mathbf{Z}$, $\{1, 7\}$ is a subgroup H' of index 2. Put $\theta(x) = 1$ for $x \in H'$ and $\theta(x) = -1$ for $x \in H - H'$, so that $\theta(xy) = \theta(x)\,\theta(y)$ for $x, y \in H$. Let w be a primitive eighth root of unity in an extension of $\mathbf{F}_p$. As in Theorem 1 consider, for $a \in H$, the "Gaussian sum"

$$\tau(a) = \sum_{x \in H} \theta(x) w^{ax}. \tag{7}$$

As in Theorem 1 we see that $\tau(a) = \theta(a)\tau(1)$ and $\tau(1)^p = \tau(p)$ (identifying p with its residue class mod 8). From the definition of $\theta(x)$ we obtain

$$\begin{aligned}\tau(1) &= w - w^3 - w^5 + w^7 = (1 - w^2)(w - w^5)\\ &= w(1 - w^2)(1 - w^4) = 2w(1 - w^2),\end{aligned}$$

since $w^8 = 1$ and $w^4 = -1$. Consequently,

$$\tau(1)^2 = 4w^2(1 - 2w^2 + w^4) = -8w^4 = 8.$$

As in Theorem 1 we show that

$$\tau(1)^p = \tau(p) = \theta(p)\tau(1).$$

Next we see that

$$\theta(p) = (\tau(1)^2)^{(p-1)/2} = 8^{(p-1)/2} = \left(\frac{8}{p}\right) \text{ (Proposition 1)} = \left(\frac{2}{p}\right)^3 = \left(\frac{2}{p}\right).$$

Therefore, $\left(\frac{2}{p}\right) = \theta(p)$. Now, by a direct calculation involving $x = 1, 3, 5, 7$ (or, more easily, $x = 1, 3, -3, -1$), it can be shown that $\theta(x) = (-1)^{(x^2-1)/8}$ and that $(-1)^{(x^2-1)/8}$ depends only on the residue class of x mod 8. Q.E.D.

Example. An application of the reciprocity law. The law of quadratic reciprocity and the complementary formulas make it possible to calculate the Legendre symbol by successive reductions. Let us calculate $\left(\frac{23}{59}\right)$ without computing squares modulo 59.

$$\begin{aligned}\left(\frac{23}{59}\right) &= (-1)^{11\cdot 29}\left(\frac{59}{23}\right) = -\left(\frac{13}{23}\right) = -(-1)^{6\cdot 11}\left(\frac{23}{13}\right) = -\left(\frac{10}{13}\right)\\ &= -\left(\frac{-3}{13}\right) = -\left(\frac{-1}{13}\right)\left(\frac{3}{13}\right) = -(-1)^6\left(\frac{3}{13}\right)\\ &= -(-1)^{6\cdot 1}\left(\frac{13}{3}\right) = -\left(\frac{1}{3}\right) = -1.\end{aligned}$$

Therefore, 23 is not a square modulo 59.

5.6. The two squares theorem

We are going to apply Proposition 1 of § 4 to the field $L = \mathbf{Q}[i]$ where $i^2 = -1$. Since $-1 \equiv 3 \pmod 4$ the ring B of integers of L is $\mathbf{Z} + \mathbf{Z}i$. This ring is called the *ring of Gaussian integers*. Its discriminant is -4 (§ 3, example). If p is an odd prime number and if u is a generator of the cyclic group $\mathbf{F}_p^*$, then $-1 = u^{(p-1)/2}$. Therefore, -1 is a square in $\mathbf{F}_p$ if and only if $(p-1)/2$ is even. We have the following classification for prime numbers:

- 2 ramifies in $\mathbf{Q}[i]$;
- the primes of the form $4k + 1$ split; and
- the primes of the form $4k + 3$ remain prime.

The following result will prove useful.

Proposition 1. The ring $B = \mathbf{Z} + \mathbf{Z}i$ *of Gaussian integers is principal.*

Proof. We use the full force of an earlier result to prove this easy proposition. With the notations of Chapter IV, § 3, we have $n = 2$, $r_1 = 0$, $r_2 = 1$, and $d = -4$. Therefore (Chapter IV, § 3, the corollary of Proposition 1),

every ideal class of B contains an integral ideal of norm $\leq \frac{4}{\pi}\frac{2}{4}|4|^{1/2} = \frac{4}{\pi}$.

Therefore, every ideal class contains the unit ideal, B itself, which is the only integral ideal of norm 1 (note that $4/\pi < 2$). Thus every ideal of B is equivalent to the principal ideal $B \cdot 1$, so B is a principal ideal ring. Q.E.D.

A sketch of an elementary proof. The points of B may be identified with $\mathbf{Z}^2 \subset \mathbf{R}^2 = \mathbf{C}$. With the usual identification of $\mathbf{R}^2$ with the plane (the square of Euclidean distance is the norm in $\mathbf{Q}[i]$) the points of $\mathbf{Z}^2$ become the vertices of squares covering the plane. A little geometry shows that, for any $x \in \mathbf{Q}[i]$, there exists $z \in B$ such that $N(x - z) = |x - z|^2 \leq (1/\sqrt{2})^2 = \frac{1}{2} < 1$. Now in $\mathfrak{a}$, a non-zero ideal of B, choose a non-zero element u of minimum norm (NB. the norm is a positive integer-valued function on $\mathbf{Z}[i] - (0)$). For any $v \in \mathfrak{a}$ there is a $z \in B$ such that $N(v/u - z) < 1$. Therefore, $N(v - zu) < N(u)$, so $v - zu = 0$ (since $v - zu \in \mathfrak{a}$). Consequently, $v \in Bu$ and $\mathfrak{a} = Bu$. The reader will observe the analogy with the proof that $\mathbf{Z}$ is principal (Chapter I, § 1).

Proposition 2 (Fermat). *Any prime number* $p \equiv 1 \pmod 4$ *may be represented as the sum of two squares (i.e. is of the form* $p = a^2 + b^2$ *with* $a, b \in \mathbf{N}$).

Proof. In fact $Bp = \mathfrak{p}_1\mathfrak{p}_2$, a product of distinct prime ideals. Clearly $p^2 = N(Bp) = N(\mathfrak{p}_1)N(\mathfrak{p}_2)$ (by Chapter III, § 5, Proposition 2). As the norms of $\mathfrak{p}_1$ and $\mathfrak{p}_2$ are larger than 1, necessarily $N(\mathfrak{p}_1) = N(\mathfrak{p}_2) = p$. Now, $\mathfrak{p}_1$ is a principal ideal $B \cdot (a + bi)$ $(a, b \in \mathbf{Z})$ (Proposition 1), so $p = N(a + bi) = a^2 + b^2$. Q.E.D.

Theorem 1. Let x *be a natural number and let* $x = \prod_p p^{v_p(x)}$ *be its expression as a product of powers of primes. For* x *to be a sum of two squares it is necessary and sufficient that, for every* $p \equiv 3 \pmod 4$, *the exponent* $v_p(x)$ *be even.*

Proof. In order to prove sufficiency we note that a sum of two squares $a^2 + b^2$ is the norm $N(a + bi)$ of an element of B. By the multiplicativity of norms, the set S of sums of two squares is itself stable under multiplication. Since $2 = 1^2 + 1^2 \in S$ and since any square is also an element of $S(x^2 = x^2 + 0^2)$, sufficiency follows from Proposition 2.

Conversely let $x = a^2 + b^2 = (a + bi)(a - bi)$ be a sum of two squares $(a, b \in \mathbf{N})$ and let p be a prime number $\equiv 3 (\mathrm{mod}\ 4)$. We have seen that the ideal Bp of B is prime (Proposition 1, § 4). Let n be the exponent of Bp in the expression for $B\cdot(a + bi)$ as a product of primes. As Bp is stable under the automorphism $\sigma: u + iv \mapsto u - iv$ of B and since $\sigma(a + bi) = a - bi$, the exponent of Bp in the expression for $B\cdot(a - bi)$ is also n. Therefore, in the expression for $B\cdot(a^2 + b^2)$, the exponent for Bp is $2n$. As no prime number besides p belongs to Bp (for $Bp \cap \mathbf{Z} = p\mathbf{Z}$), we see that $v_p(x) = 2n$ and $v_p(x)$ is even. Q.E.D.

5.7. The four squares theorem

In this section we intend to prove the following theorem:

Theorem 1 (Lagrange). *Every natural number may be represented as the sum of four squares.*

The idea behind this proof is analogous to that of § 6: in place of the ring of Gaussian integers, we will work in a suitably chosen ring of *quaternions*.

Let us begin with a definition of quaternions. Given a ring A, we will write $(1, i, j, k)$ for the canonical base of the A-module A^4 and define a multiplication on A^4 as follows:

$$(1) \qquad \begin{cases} 1 \text{ is the unit element,} \\ i^2 = j^2 = k^2 = -1, \\ ij = -ji = k, \quad jk = -kj = i, \quad \text{and} \quad ki = -ik = j. \end{cases}$$

We extend this multiplication to elements $a + bi + cj + dk$ of A^4 by linearity. Distributivity is then clear. It suffices to verify associativity for elements of the base, e.g. $i(jk) = i^2 = -1 = k^2 = (ij)k$. The formulas in which 1 appears being clear, there remain $3^3 - 1 = 26$ formulas to check. The patient and incredulous reader will reduce the number of formulas by the use of permutations and check those which remain. Others will accept the author's claim that the given multiplication is associative. Provided with this multiplication, A^4 is a not necessarily commutative ring, and even an A-algebra, which we call the *algebra of quaternions over* A, denoted by $\mathbf{H}(A)$ ($\mathbf{H}$ in honor of W. R. Hamilton, the inventor of quaternions).

Given a quaternion $z = a + bi + cj + dk$ over A (we write a in place of $a\cdot 1$), we define the *conjugate* quaternion of z to be the quaternion $\bar{z} = a - bi - cj - dk$.

Lemma 1. For any z and $z' \in \mathbf{H}(A)$, $\overline{z + z'} = \bar{z} + \overline{z'}$, $\overline{zz'} = \overline{z'}\bar{z}$, and $\bar{\bar{z}} = z$. In other words, $z \mapsto \bar{z}$ is an involutive antiautomorphism of $\mathbf{H}(A)$.

Proof. The first and the third formulas are clear. By linearity we see that to prove the second it suffices to check that $\overline{xy} = \bar{y}\bar{x}$ for $x, y \in \{1, i, j, k\}$. This is clear if $x = 1$ or $y = 1$. If $x = y = i$, then

$$\overline{xy} = -\bar{1} \quad \text{and} \quad \bar{y}\bar{x} = (-i)(-i) = i^2 = -1.$$

If $x = i$ and $y = j$, then

$$\overline{xy} = \bar{k} = -k \quad \text{and} \quad \bar{y}\bar{x} = (-j)(-i) = ji = -k.$$

The others follow by a permutation argument. Q.E.D.

Given a quaternion z over A, we call the quaternion $\mathrm{N}(z) = z\bar{z}$ the *reduced norm* of z.

Lemma 2.

(a) *For* $z = a + bi + cj + dk \in \mathbf{H}(\mathrm{A})$,
$\mathrm{N}(z) = a^2 + b^2 + c^2 + d^2$ (four squares!), *an element of* A.

(b) *For* $z, z' \in \mathbf{H}(\mathrm{A})$, $\mathrm{N}(zz') = \mathrm{N}(z)\mathrm{N}(z')$.

Proof. To prove (a) we observe that, in the expansion of $(a + bi + cj + dk)(a - bi - cj - dk)$, the cross-product terms cancel, leaving $a^2 + b^2 + c^2 + d^2$. To prove (b), observe that

$$\mathrm{N}(zz') = zz'\overline{zz'} = zz'\,\bar{z}'\bar{z} = z\mathrm{N}(z')\bar{z} = z\bar{z}\mathrm{N}(z')$$

(since $\mathrm{N}(z') \in \mathrm{A}$ and any element of A commutes with every quaternion), so $N(zz') = N(z)N(z')$. Q.E.D.

Lemma 2 implies that, for a (commutative) ring A, the set of reduced norms of quaternions, i.e. sums of four squares, is stable under multiplication.

Now we shall study in greater detail the quaternion algebra $\mathbf{H}(\mathbf{Q})$, the non-commutative subring $\mathbf{H}(\mathbf{Z})$, and the subset $\mathbf{H}$ of "quaternions of Hurwitz", i.e. the quaternions of the form $a + bi + cj + dk$, where either a, b, c, and d belong to $\mathbf{Z}$ or all four coefficients belong to $\frac{1}{2} + \mathbf{Z}$.

Lemma 3.

(a) *The set* $\mathbf{H}$ *of Hurwitz quaternions is a non-commutative subring of* $\mathbf{H}(\mathbf{Q})$ *containing* $\mathbf{H}(\mathbf{Z})$ *and stable under conjugation* $(z \mapsto \bar{z})$.

(b) *For any* $z \in \mathbf{H}$, $z + \bar{z} \in \mathbf{Z}$ *and* $\mathrm{N}(z) = z\bar{z} \in \mathbf{Z}$.

(c) *In order that* $z \in \mathbf{H}$ *be invertible, it is necessary and sufficient that* $\mathrm{N}(z) = 1$.

(d) *Every left ideal (respectively, right ideal)* $\mathfrak{a}$ *of* $\mathbf{H}$ *is principal, i.e. of the form* $\mathbf{H}z$ *(respectively,* $z\mathbf{H}$*).*

Proof. In (a) all assertions are clear except the stability of $\mathbf{H}$ under multiplication. To complete the proof of (a) it suffices to show that, for $u = \frac{1}{2}(1 + i + j + k)$, $u \cdot 1$, $u \cdot i$, $u \cdot j$, $u \cdot k$, and $u^2 \in \mathbf{H}$. We have

$$u \cdot 1 = u, \quad u \cdot i = \tfrac{1}{2}(-1 + i + j - k), \quad u \cdot j = \tfrac{1}{2}(-1 - i + j + k),$$
$$\text{and} \quad u \cdot k = \tfrac{1}{2}(-1 + i - j + k),$$

all elements of $\mathbf{H}$. By addition, $2u^2 = \frac{1}{2}(-2 + 2i + 2j + 2k)$, so $u^2 \in \mathbf{H}$, too. To prove (b) let

$$z = \tfrac{1}{2} + a + (\tfrac{1}{2} + b)i + (\tfrac{1}{2} + c)j + (\tfrac{1}{2} + d)k, \quad \text{with } a, b, c, d \in \mathbf{Z}.$$

Then

$$z + \bar{z} = 1 + 2a \in \mathbf{Z}$$

and

$$z\bar{z} = (\tfrac{1}{2} + a)^2 + (\tfrac{1}{2} + b)^2 + (\tfrac{1}{2} + c)^2 + (\tfrac{1}{2} + d)^2 \in \tfrac{4}{4} + \mathbf{Z} \subset \mathbf{Z}.$$

The preceding formula follows from Lemma 2.

If z is invertible in $\mathbf{H}$ and if $z' = z^{-1}$, then

$$\mathrm{N}(z)\mathrm{N}(z') = \mathrm{N}(zz') = 1.$$

Since $\mathrm{N}(z)$ and $\mathrm{N}(z')$ are both positive integers ((b) and Lemma 2, (a)), $\mathrm{N}(z) = 1$. Conversely, if $z \in \mathbf{H}$ and if $\mathrm{N}(z) = 1$, then

$$z\bar{z} = \bar{z}z = \mathrm{N}(z) = 1,$$

so, since $\bar{z} \in \mathbf{H}$ by (a), z is invertible in $\mathbf{H}$. This proves (c).

To prove (d) take $x = a + bi + cj + dk \in \mathbf{H}(\mathbf{Q})$. There exist $a', b', c', d' \in \mathbf{Z}$ such that

$$|a - a'| \leq \tfrac{1}{2}, \quad |b - b'| \leq \tfrac{1}{2}, \quad |c - c'| \leq \tfrac{1}{2}, \quad \text{and} \quad |d - d'| \leq \tfrac{1}{2}.$$

Put $z = a' + b'i + c'j + d'k$. Then

$$\mathrm{N}(x - z) = (a - a')^2 + (b - b')^2 + (c - c')^2 + (d - d')^2 \leq 4 \cdot \tfrac{1}{4} = 1.$$

The inequality is even strict, except when a, b, c, and d all belong to $\frac{1}{2} + \mathbf{Z}$. But in this case $x \in \mathbf{H}$. Thus, for any quaternion $x \in \mathbf{H}(\mathbf{Q})$, there exists a Hurwitz quaternion $z \in \mathbf{H}$ such that $\mathrm{N}(x - z) < 1$ (it is precisely because we must have strict inequality that we work with the Hurwitz quaternions; $\mathbf{H}(\mathbf{Z})$ would not be sufficient). Now let $\mathfrak{a}$ be a left ideal of $\mathbf{H}$. To show that $\mathfrak{a}$ is principal we may assume $\mathfrak{a} \neq (0)$. Choose in $\mathfrak{a}$ a non-zero element u for which $\mathrm{N}(u)$ is a minimum (such u exists, since the reduced norm is a positive integer-valued function on $\mathbf{H} - (0)$, by (b)). Clearly, u is invertible in $\mathbf{H}(\mathbf{Q})$ with inverse $\bar{u}\mathrm{N}(u)^{-1}$ (this shows that $\mathbf{H}(\mathbf{Q})$ is a skew field). For $y \in \mathfrak{a}$ form $yu^{-1} \in \mathbf{H}(\mathbf{Q})$ and take an element $z \in \mathbf{H}$ such that $\mathrm{N}(yu^{-1} - z) < 1$. Then, by Lemma 2, (b), we have

$$\mathrm{N}(y - zu) = \mathrm{N}((yu^{-1} - z)u) < \mathrm{N}(u).$$

Since $y - zu \in \mathfrak{a}$ and since $\mathrm{N}(u)$ is as small as possible, it follows that $y - zu = 0$, $y \in \mathbf{H}u$, and $\mathfrak{a} = \mathbf{H}u$. Q.E.D.

Now we are ready to prove Theorem 1. Since the set of elements of $\mathbf{Z}$ which are sums of four squares is multiplicatively stable (cf. Lemma 2), Theorem 1 follows from the following proposition.

Proposition 1. Any prime number is the sum of four squares.

Proof. Since $2 = 1^2 + 1^2 + 0^2 + 0^2$, we may suppose that p is odd. Since p commutes with any quaternion, the left ideal $\mathbf{H}p$ is two-sided. Consider the quotient ring $\mathbf{H}/\mathbf{H}p$. Since p is odd, any $z \in \mathbf{H}$ is congruent mod $\mathbf{H}p$ to an element of $\mathbf{H}(\mathbf{Z})$ (if the components of z all belong to $\frac{1}{2} + \mathbf{Z}$, form $z + p \cdot \frac{1}{2}(1 + i + j + k)$). Therefore, $\mathbf{H}/\mathbf{H}p$ is isomorphic to the corresponding quotient ring of $\mathbf{H}(\mathbf{Z})$, i.e. $\mathbf{H}(\mathbf{F}_p)$.

Since the quadratic form $a^2 + b^2 + c^2 + d^2$ represents 0 in $\mathbf{F}_p$ (Chapter I, § 7, Theorem 2; see the remark below for a direct proof), $\mathbf{H}(\mathbf{F}_p)$ contains a non-zero element of reduced norm zero. Such an element is not invertible (Lemma 2, (b)), so it generates a non-trivial left ideal. It follows that $\mathbf{H}p$ is properly contained in a left ideal $\mathbf{H}z$ distinct from $\mathbf{H}$. Consequently, $p = z'z$ for some $z, z' \in \mathbf{H}$ both non-units. Thus,

$$p^2 = \mathrm{N}(p) = \mathrm{N}(z)\,\mathrm{N}(z')$$

and, since $\mathrm{N}(z)$ and $\mathrm{N}(z')$ are both integers strictly larger than one (Lemma 3, (b) and (c)), $\mathrm{N}(z) = \mathrm{N}(z') = p$.

Put $z = a + bi + cj + dk$ ($a, b, c, d \in \mathbf{Z}$ or in $\frac{1}{2} + \mathbf{Z}$). If $a, b, c, d \in \mathbf{Z}$, then $p = \mathrm{N}(z) = a^2 + b^2 + c^2 + d^2$ and we are finished. It suffices to show that, if $a, b, c, d \in \frac{1}{2} + \mathbf{Z}$, we may make a reduction to the preceding case by multiplying z by an element of $\mathbf{H}$ of reduced norm 1, more precisely by an element of the form $\frac{1}{2}(\pm 1 \pm i \pm j \pm k)$. To see this consider the residue class η associated with $2z$ in $\mathbf{H}(\mathbf{Z})/4\mathbf{H}(\mathbf{Z}) \cong \mathbf{H}(\mathbf{Z}/4\mathbf{Z})$. Since $\mathrm{N}(z) \in \mathbf{Z}$, $\mathrm{N}(2z) \in 4\mathbf{Z}$, so $\mathrm{N}(\eta) = 0$ and $\eta\bar{\eta} = 0$. We observe that $\bar{\eta}$ is the residue class of a quaternion z' of the form $\pm 1 \pm i \pm j \pm k$. Set $u = \frac{1}{2}z' \in \mathbf{H}$; then u is of reduced norm 1 and, inasmuch as $(2z)(2u)$ is zero mod 4, $zu \in \mathbf{H}(\mathbf{Z})$. Now $p = \mathrm{N}(z) = \mathrm{N}(zu)$, so we are finished. Q.E.D.

Remark. Here is a very elementary proof of the fact that, over a finite field K, the quadratic form $a^2 + b^2 + c^2 + d^2$ represents 0 (i.e. has a non-trivial zero in K^4). It suffices to take $c = 1$ and $d = 0$ and to show that the equation $a^2 + b^2 + 1 = 0$ has a solution in K^2. Write this equation in the form $b^2 + 1 = -a^2$. In characteristic 2, we may take $b = 0$ and $a = 1$. If $\mathrm{card}(\mathrm{K}) = q$ is odd, there are $(q + 1)/2$ squares in K (0 and the $(q - 1)/2$ non-zero squares). Thus the set T (respectively, T′) of elements of K of the form $b^2 + 1$ with $b \in \mathrm{K}$ (respectively, of the form $-a^2$ with $a \in \mathrm{K}$) consists of $(q + 1)/2$ elements. Since $(q + 1)/2 + (q + 1)/2 > q$, $\mathrm{T} \cap \mathrm{T}' \neq \phi$, so $b^2 + 1 = -a^2$ has a solution. Q.E.D.

Chapter VI

Galois extensions of number fields

6.1. Galois theory

This section is a supplement to the general theory of commutative fields presented in Chapter II, §§ 3, 4, 6, and 7. Given a field L and a set G of automorphisms of L, one sees immediately that the set of all $x \in L$ such that $\sigma(x) = x$ for all $\sigma \in G$ is a subfield of L, called the *fixed field* of G. It is also clear that, for an extension L of a field K, the set of K-automorphisms of L is a group under composition of mappings.

Theorem 1. Let L *be an extension of finite degree n of a field* K, *where* K *is finite or of characteristic zero. The following conditions are equivalent:*

(a) K *is the fixed field of the group* G *of* K*-automorphisms of* L.
(b) *For every* $x \in L$, *the minimal polynomial of* x *over* K *has all its roots in* L.
(c) L *is generated by the roots of a polynomial with coefficients in* K.

Under the above conditions the group G *of* K*-automorphisms of* L *is of order* n.

Proof. To see that (a) implies (b) observe that, for $x \in L$, the polynomial $\prod_{\sigma \in G} (X - \sigma(x)) = P(X)$ is invariant under G (since any element of G permutes the factors).[1] Therefore, the coefficients of P(X) belong to K. Since x is a root of P(X) (G contains an identity element), the minimal polynomial of x divides P(X) (Chapter II, § 3, (4)). Therefore, (a) implies (b).

To prove that (b) implies (c) take a primitive element x of L over K (Chapter II, § 4, corollary of Theorem 1). Its minimal polynomial over K has all its roots in L by (b). Clearly, these roots generate L over K (since any one of them does).

Finally, let us show that (c) implies (a). By hypothesis L is generated over K by a finite number of elements ($x^{(1)}, \ldots, x^{(q)}$) and by all their conjugates ($x_j^{(i)}$) (Chapter II, § 4). It is clear that any K-isomorphism σ of L into an extension of L sends each of these generators to another element of the same set. Therefore, $\sigma(L) \subset L$. Moreover, by linear algebra, $\sigma(L) = L$, since σ is K-linear and injective. In other words, σ is a K-*automorphism* of L. It follows from Chapter II, § 4, Theorem 1 that the group G of K-automorphisms of L has precisely n elements. Let $x \in L$ be invariant under G. Then every $\sigma \in G$ is a K[x]-automorphism of L. By Chapter II, § 4, Theorem 1 there are exactly

1. The finiteness of G follows from Chapter II, § 4, Theorem 1.

$[L:K[x]]$ $K[x]$-isomorphisms of L into an extension field of L. Thus, $n \leq [L:K[x]]$, from which we may conclude that $n = [L:K[x]]$, $K[x] = K$, and $x \in K$. This proves that (c) implies (a). That the order of G is n has been proved along the way. Q.E.D.

Definition 1. If the conditions of Theorem 1 are satisfied, L *is called a Galois extension of* K *and* G *is called the Galois group of* L *over* K. *If* G *is abelian (respectively, cyclic),* L *is called an abelian (respectively, cyclic) extension of* K.

Corollary of Theorem 1. Let K *be a finite field or a field of characteristic zero, let* L *be an extension of finite degree n of* K, *and let* H *be a group of automorphisms of* L *such that* K *is the fixed field of* H. *Then* L *is a Galois extension of* K *and* H *is the Galois group of* L *over* K.

Proof. For $x \in L$ the polynomial $\prod_{\sigma \in H} (X - \sigma(x)) = P(X)$ is invariant under H. Therefore, $P(X)$ has its coefficients in K. Since x is a root of $P(X)$, the minimal polynomial of x over K divides $P(X)$. By Theorem 1, (b) L is a Galois extension of K. Let G be the Galois group of L over K. We have $H \subset G$ and $\text{card}(G) = n$ (Theorem 1). Now let x be a primitive element of L over K (Chapter II, § 4, corollary of Theorem 1) and let $P(X)$ be the polynomial constructed above. Since $n \leq d°(P)$ and $d°(P) = \text{card}(H) \leq \text{card}(G) = n$, we see that $H = G$. Q.E.D.

Theorem 2. Let K *be a field which is finite or of characteristic zero, let* L *be a Galois extension of* K, *and let* G *be the Galois group of* L *over* K. *To any subgroup* G′ *of* G *let us associate the fixed field* $k(G')$ *of* G′, *and to any subfield* K′ *of* L *containing* K *let us associate the subgroup* $g(K')$ *of* G *consisting of all* K′*-automorphisms of* L.

(a) *The mappings g and k are bijections and are inverses of one another. They are both decreasing with respect to the inclusion relations on* G *and on* L, *i.e. they reverse inclusions. The field* L *is a Galois extension of any intermediate field* K′ (i.e. $K \subset K' \subset L$).
(b) *In order that an intermediate field* K′ *be a Galois extension of* K, *it is necessary and sufficient that* $g(K')$ *be a normal subgroup of* G. *In this case the Galois group of* K′ *over* K *may be identified with the quotient group* $G/g(K')$.

Proof. For any intermediate field K′ and any $x \in L$ the minimal polynomial of x over K′ divides the minimal polynomial of x over K. Thus, all its roots belong to L by Theorem 1, (b), so L is a Galois extension of K′, also by Theorem 1, (b). K′ is the fixed field of the group $g(K')$ of all K′-automorphisms of L (Theorem 1, (a)); in other words, $k(g(K')) = K'$. Let G′ be a subgroup of G. Then G′ is the Galois group of L over $k(G')$ (corollary of Theorem 1); this means $G' = g(k(G'))$. The relations $k(g(K')) = K'$ and $g(k(G')) = G'$ imply that k and g are bijections and inverses of one another. It is clear that they reverse inclusions. This proves (a).

Let us prove (b). Let K′ be an intermediate field ($K \subset K' \subset L$). For $x \in K'$ the roots of the minimal polynomial of x over K are the elements of L of the form $\sigma(x)$, for $\sigma \in G$. According to Theorem 1, (b), in order that K′ be a Galois extension of K, it is necessary and sufficient that $\sigma(x) \in K'$ for all $x \in K'$ and all $\sigma \in G$, i.e. that $\sigma(K') \subset K'$ for all $\sigma \in G$. But, if $\sigma(K') \subset K'$, if $\tau \in g(K')$, and if $x \in K'$, then $\sigma^{-1}\tau\sigma(x) = \sigma^{-1}\sigma(x) = x$, so $\sigma^{-1}\tau\sigma \in g(K')$. In other words, K′ Galois over K implies $g(K')$ normal in G. Conversely, suppose that $g(K')$ is normal in G. If $x \in K'$, $\sigma \in G$, and $\tau \in g(K')$, then $\tau\sigma(x) = \sigma\sigma^{-1}\tau\sigma(x) = \sigma(x)$, since $\sigma^{-1}\tau\sigma \in g(K')$. Thus $\sigma(x)$ is invariant under every element of $g(K')$, so $\sigma(x) \in K'$. Consequently, $g(K')$ normal in G implies $\sigma(K') \subset K'$ and so K' is Galois over K.

Finally, let us determine the Galois group of K' over K. Since $\sigma(K') \subset K'$ for all $\sigma \in G$ (and even $\sigma(K') = K'$ by linear algebra), the restriction $\sigma \mid K'$ of σ to K' is a K-automorphism of K'. Restriction, $\sigma \mapsto \sigma \mid K'$, is a homomorphism of G to the Galois group H of K' over K. Clearly, its kernel is $g(K')$. Since $\text{card}(H) = [K' : K] = [L : K][L : K']^{-1} = \text{card}(G) \cdot \text{card}(g(K'))^{-1} = \text{card}(G/g(K'))$, we may conclude that the restriction homomorphism is surjective. Therefore, $H \cong G/g(K')$. Q.E.D.

Example 1. Quadratic extensions. Let K be a field of characteristic zero, and let L be a quadratic extension (i.e. of degree 2) of K. As in the beginning of § 5, Chapter II one sees that L is of the form $K[x]$, where x is a root of a polynomial $X^2 - d$ ($d \in K$ and d non-square). Since the other root of this polynomial is $-x$, there is a non-trivial K-automorphism σ defined by $\sigma(x) = -x$, i.e. $\sigma(a + bx) = a - bx (a, b \in K)$. Clearly $\sigma^2 = 1$ and K is the fixed field for σ. Thus L is a Galois extension of K with the cyclic Galois group $\{1, \sigma\}$ (by Theorem 1 and its corollary).

Example 2. Cyclotomic extensions. Let K be a field of characteristic zero, let z be a primitive nth root of unity in an extension of K, and let $L = K(z)$. The field L is called a *cyclotomic* extension of K. The minimal polynomial of z over K divides $X^n - 1$ (Chapter II, § 3, (4)), so its roots are nth roots of unity and, consequently, powers of z (Chapter I, § 6). Thus L is a Galois extension of K by Theorem 1, (c).

Let G be the Galois group of L over K. Any $\sigma \in G$ is determined by its effect on z. More precisely $\sigma(z)$ is a power $z^{j(\sigma)}$ of z where $j(\sigma)$ is uniquely determined modulo n. For $\sigma, \tau \in G$ one sees that $\sigma\tau(z) = \sigma(z^{j(\tau)}) = \sigma(z)^{j(\tau)} = z^{j(\sigma)j(\tau)}$, so $j(\sigma\tau) \equiv j(\sigma)j(\tau)$ (modulo n). In other words $\sigma \mapsto j(\sigma)$ defines a homomorphism $G \to (\mathbf{Z}/n\mathbf{Z})^*$. Since $j(\sigma)$ determines σ, this homomorphism is injective, and G is abelian. Thus *any cyclotomic extension is abelian.* If n is prime, this extension is even *cyclic*, for G is isomorphic to a subgroup of $(\mathbf{Z}/n\mathbf{Z})^* = \mathbf{F}_n^*$ (Chapter I, § 7, Theorem 1, (b)).

As any subgroup of an abelian group is normal, any intermediate field K' in a cyclotomic extension is a Galois extension (and an abelian extension) of K (Theorem 2, (b)). In particular, any subfield of a cyclotomic field is an abelian extension of $\mathbf{Q}$. Conversely, it can be shown (the theorem of Kronecker and Weber) that any abelian extension of $\mathbf{Q}$ is a subfield of a cyclotomic field.

The reader will observe that, with the preceding notations, the automorphism σ raises every nth root of unity to its $j(\sigma)$ power, for all the nth roots of unity are powers of z. Thus $\sigma \mapsto j(\sigma)$ is independent of the choice of z.

Example 3. Finite fields. Let $\mathbf{F}_q$ be a finite field ($q = p^s$ with p prime). Any extension of finite degree of $\mathbf{F}_q$ is of the form $\mathbf{F}_{q^n}$. Its degree is n (Chapter I, § 7). The mapping $\sigma: x \mapsto x^q$ is an automorphism of $\mathbf{F}_{q^n}$ (ibid., Proposition 1) with $\mathbf{F}_q$ as its fixed field (ibid., Theorem 1, (c)). For any $x \in \mathbf{F}_{q^n}$ we have $\sigma^j(x) = x^{q^j}$ and $\sigma^n = 1$, since $x \in \mathbf{F}_{q^n}$ satisfies the relation $x^{q^n} = x$ (ibid., Theorem 1, (c)). On the other hand, for $1 \leq j \leq n - 1$, $\sigma^j \neq 1$ since, if $j < n$, there exists $x \in \mathbf{F}_{q^n}$ such that $x^{q^j} \neq x$. Thus, $\{1, \sigma, \ldots, \sigma^{n-1}\}$ is a cyclic group of order n. According to the corollary of Theorem 1, $\mathbf{F}_{q^n}$ *is a cyclic extension of degree n of* $\mathbf{F}_q$. *Its Galois group has a canonical generator, the mapping* $x \mapsto x^q$. *This mapping is called the Frobenius automorphism.*

6.2. The decomposition and inertia groups

In this section A *denotes a Dedekind ring,* K *its field of fractions (which is assumed of characteristic zero),* K′ *is a Galois extension of* K *of degree* n, G *is the Galois group of* K′ *over* K, *and* A′ *is the integral closure of* A *in* K′.

By applying $\sigma \in G$ to an equation of integral dependence (over A) of an element of A′, one sees that $\sigma(x) \in A'$. Therefore,

(1) A′ *is stable under* G, *i.e.* $\sigma(A') = A'$ *for all* $\sigma \in G$.

We have shown only that $\sigma(A') \subset A'$. But we also have $\sigma^{-1}(A') \subset A'$, so $A' = \sigma\sigma^{-1}(A') \subset \sigma(A')$. From now on we shall leave the details of arguments of this sort to the reader.

On the other hand, if $\mathfrak{p}$ is a maximal ideal of A and $\mathfrak{p}'$ a maximal ideal of A′ such that $\mathfrak{p}' \cap A = \mathfrak{p}$ (i.e. $\mathfrak{p}'$ appears in the factorization of $A'\mathfrak{p}$ into a product of prime ideals in A′; cf. Chapter V, § 2, Proposition 1), then, clearly $\sigma(\mathfrak{p}') \cap A = \mathfrak{p}$, so $\sigma(\mathfrak{p}')$ also appears in the expression for $A'\mathfrak{p}$ as a product of prime ideals of A′ and with the same exponent as $\mathfrak{p}'$. We shall say that $\mathfrak{p}'$ and $\sigma(\mathfrak{p}')$ are *conjugate* prime ideals of A′. We are going to show that all the prime ideals in the prime factorization of $A'\mathfrak{p}$ in A′ are conjugate.

Proposition 1. If $\mathfrak{p}$ *is a maximal ideal of* A, *then the maximal ideals* $\mathfrak{p}_i'$ *of* A′ *which appear in the expression for* $A'\mathfrak{p}$ *as a product of prime ideals in* A′ *(i.e. the maximal ideals* $\mathfrak{p}_i'$ *are characterized by the property* $\mathfrak{p}_i' \cap A = \mathfrak{p}$*) are all conjugate. They have the same residual degree* f *and the same ramification index* e. *Thus* $A'\mathfrak{p} = (\prod_{i=1}^{g} \mathfrak{p}_i')^e$ *and* $n = efg$.

Proof. The assertion concerning the ramification index and the residual degree is obvious, because an automorphism preserves *all* algebraic relations. The formula $n = efg$ is thus a special case of the relation $\sum e_i f_i = n$ (Chapter V, § 2, Theorem 1). Now let $\mathfrak{p}'$ be one of the $\mathfrak{p}_i'$'s and assume that another of the $\mathfrak{p}_i'$'s, which we shall denote $\mathfrak{q}'$, is not conjugate to $\mathfrak{p}'$. Since $\mathfrak{q}'$ and $\sigma(\mathfrak{p}')$ $(\sigma \in G)$ are maximal and distinct, $\sigma(\mathfrak{p}') \not\subset \mathfrak{q}'$. We need the following lemma.

Lemma 1 ("lemma for staying away from prime ideals"). *Let* R *be a ring,* $\mathfrak{p}_1, \ldots, \mathfrak{p}_q$ *a finite set of prime ideals of* R, *and let* $\mathfrak{b}$ *be an ideal of* R *such that* $\mathfrak{b} \not\subset \mathfrak{p}_i$ *for any index* i. *Then there exists* $b \in \mathfrak{b}$ *such that* $b \notin \mathfrak{p}_i$ *for any* i.

Proof of the Lemma. Without loss of generality we may assume $\mathfrak{p}_j \not\subset \mathfrak{p}_i$, $i \neq j$. Let $x_{ij} \in \mathfrak{p}_j - \mathfrak{p}_i$ for $i \neq j$, $1 \leq i, j \leq q$. Since $\mathfrak{b} \not\subset \mathfrak{p}_i$, there exists $a_i \in \mathfrak{b} - \mathfrak{p}_i$. Put $b_i = a_i \prod_{j \neq i} x_{ij}$. Then $b_i \in \mathfrak{b}$, $b_i \in \mathfrak{p}_j$ for $j \neq i$, and $b_i \notin \mathfrak{p}_i$ (since $\mathfrak{p}_i$ is prime). Thus $b = b_1 + \cdots + b_q \in \mathfrak{b} - \bigcup_{i=1}^{q} \mathfrak{p}_i$, since clearly $b \in \mathfrak{b}$ and $b = b_i$ (modulo $\mathfrak{p}_i$) (so $b \notin \mathfrak{p}_i$ for any i).

Q.E.D.

Returning to the proof of the proposition, we see, from the lemma, that there is an element $x \in \mathfrak{q}'$ such that $x \notin \sigma(\mathfrak{p}')$ for all $\sigma \in G$. Consider the norm of x, $N(x) = \prod_{\tau \in G} \tau(x)$ (Chapter II, § 6, Proposition 1). Since $\tau(x) \in A'$ for all $\tau \in G$ (by (1)) we see that

$N(x) \in \mathfrak{q}'$; in fact $N(x) \in \mathfrak{q}' \cap A = \mathfrak{p}$. On the other hand $x \notin \tau^{-1}(\mathfrak{p}')$, so $\tau(x) \notin \mathfrak{p}'$ for any $\tau \in G$. Since $\mathfrak{p}'$ is prime, we see that $N(x) \notin \mathfrak{p}'$, and this contradicts $N(x) \in \mathfrak{p}$. Q.E.D.

Now let $\mathfrak{p}'$ be a maximal ideal of A' such that $\mathfrak{p}' \cap A = \mathfrak{p}$. Those $\sigma \in G$ for which $\sigma(\mathfrak{p}') = \mathfrak{p}'$ form a subgroup D of G, called the *decomposition group of* $\mathfrak{p}'$. If g denotes the number of conjugates of $\mathfrak{p}'$, then

$$g = \text{card}(G)\cdot\text{card}(D)^{-1} \quad \text{or} \quad \text{card}(D) = \frac{n}{g} = ef. \tag{2}$$

For $\sigma \in D$ the relations $\sigma(A') = A'$ and $\sigma(\mathfrak{p}') = \mathfrak{p}'$ imply that σ induces an automorphism $\bar{\sigma}$ of $A'/\mathfrak{p}'$ ($x \equiv y \pmod{\mathfrak{p}'}$ entails $\sigma(x) \equiv \sigma(y) \pmod{\mathfrak{p}'}$). It is clear that $\bar{\sigma}$ is an $(A/\mathfrak{p})$-automorphism. The mapping $\sigma \mapsto \bar{\sigma}$ is a group homomorphism, whose kernel is the sub-group $I \subset D$ consisting of those $\sigma \in D$ which satisfy the relation $\sigma(x) - x \in \mathfrak{p}'$ for all $x \in A'$. Thus, I is a normal subgroup of D, called the *inertia subgroup of* $\mathfrak{p}'$.

Proposition 2. With the same notations as above, assume that $A/\mathfrak{p}$ *is finite or of characteristic zero. Then* $A'/\mathfrak{p}'$ *is a Galois extension of degree* f *of* $A/\mathfrak{p}$, *and the mapping* $\sigma \mapsto \bar{\sigma}$ *is a surjective homomorphism of* D *on the Galois group of* $A'/\mathfrak{p}'$ *over* $A/\mathfrak{p}$. *Moreover,* $\text{card}(I) = e$.

Proof. Let K_D be the fixed field of D, let $A_D = A' \cap K_D$ be the integral closure of A in K_D, and let $\mathfrak{p}_D$ be the prime ideal $\mathfrak{p}' \cap A_D$. According to Proposition 1 and the definition of D, $\mathfrak{p}'$ is the only prime factor of $A'\mathfrak{p}_D$. Put $A'\mathfrak{p}_D = \mathfrak{p}'^{e'}$ and write f' for the residual degree $[A'/\mathfrak{p}' : A_D/\mathfrak{p}_D]$. According to Theorem 1 of § 2, Chapter V, Theorem 2 of § 1, and (2), we have $e'f' = [K' : K_D] = \text{card}(D) = ef$. Since $A/\mathfrak{p} \subset A_D/\mathfrak{p}_D \subset A'/\mathfrak{p}'$, $f' \leq f$. Since $\mathfrak{p}A_D \subset \mathfrak{p}_D$, $e' \leq e$. Therefore, since $e'f' = ef$, $e = e'$ and $f = f'$. This entails

$$A/\mathfrak{p} \cong A_D/\mathfrak{p}_D. \tag{3}$$

Now let $\bar{x}$ be a primitive element for $A'/\mathfrak{p}'$ over $A/\mathfrak{p}$ and let $x \in A'$ be a representative for $\bar{x}$. Let $X^r + a_{r-1}X^{r-1} + \cdots + a_0 = P(X)$ be the minimal polynomial for x over K_D. We know that $a_i \in A_D$ (Chapter II, § 6, the corollary of Proposition 2). The roots of $P(X)$ are all of the form $\sigma(x)$ with $\sigma \in D$. The "reduced" polynomial $\bar{P}(X) = X^r + \bar{a}_{r-1}X^{r-1} + \cdots + \bar{a}_0$ has its coefficients in $A/\mathfrak{p}$ (by (3)) and the roots of $\bar{P}(X)$ are all of the form $\bar{\sigma}(\bar{x})$ with $\sigma \in D$. Consequently, $A'/\mathfrak{p}'$ contains all the conjugates of $\bar{x}$ over $A/\mathfrak{p}$, and $A'/\mathfrak{p}'$ is a Galois extension of $A/\mathfrak{p}$ (§ 1, Theorem 1, (c)). Furthermore, since every conjugate of $\bar{x}$ over $A/\mathfrak{p}$ is of the form $\bar{\sigma}(\bar{x})$, every $(A/\mathfrak{p})$-automorphism of $A'/\mathfrak{p}'$ is of the form $\bar{\sigma}$ for some $\sigma \in D$. Thus, the Galois group of $A'/\mathfrak{p}'$ over $A/\mathfrak{p}$ may be identified with D/I. Since its order is $[A'/\mathfrak{p}' : A/\mathfrak{p}] = f$, $\text{card}(D)/\text{card}(I) = f$, so $\text{card}(I) = e$, by (2). Q.E.D.

Corollary. In order that $\mathfrak{p}$ *not ramify in* A' *it is necessary and sufficient that the inertia group* I *of* $\mathfrak{p}'$ *for any* $\mathfrak{p}'$ *over* $\mathfrak{p}$ *(i.e. such that* $\mathfrak{p}' \cap A = \mathfrak{p}$*) be trivial.*

Remark. Write $D_{\mathfrak{p}'}$ and $I_{\mathfrak{p}'}$ for the decomposition and inertia groups of the maximal ideal $\mathfrak{p}' \subset A'$. For a conjugate ideal $\sigma(\mathfrak{p}')$ ($\sigma \in G$)

$$D_{\sigma(\mathfrak{p}')} = \sigma D_{\mathfrak{p}'}\sigma^{-1} \quad \text{and} \quad I_{\sigma(\mathfrak{p}')} = \sigma I_{\mathfrak{p}'}\sigma^{-1}. \tag{4}$$

To prove (4) note that, for $\tau \in D_{\mathfrak{p}'}$, we have

$$\sigma\tau\sigma^{-1}\cdot\sigma(\mathfrak{p}') = \sigma\tau(\mathfrak{p}') = \sigma(\mathfrak{p}'),$$

so $\sigma D_{\mathfrak{p}'}\sigma^{-1} \subset D_{\sigma(\mathfrak{p}')}$.
The reverse inclusion follows by an argument analogous to that of the remark following (1). Similarly, for $\tau \in I_{\mathfrak{p}'}$ and $x \in A'$,

$$\sigma\tau\sigma^{-1}(x) - x = \sigma\tau(\sigma^{-1}(x)) - \sigma\sigma^{-1}(x) = \sigma(\tau(\sigma^{-1}(x)) - \sigma^{-1}(x)) \in \sigma(\mathfrak{p}'),$$

so $\sigma I_{\mathfrak{p}'}\sigma^{-1} \subset I_{\sigma(\mathfrak{p}')}$, the reverse inclusion being proved as before.

When K′ is an *abelian* extension of K, the groups $D_{\sigma(\mathfrak{p}')}$ (respectively, $I_{\sigma(\mathfrak{p}')}$) $(\sigma \in G)$ are all *equal*. They depend only on the ideal $\mathfrak{p}$ of the ring A (Proposition 1).

6.3. The number field case. The Frobenius automorphism

The preceding will now be applied to number fields and their rings of integers. This application is possible because number fields are of characteristic zero and their residual fields are finite.

Let us keep our earlier notations ($K \subset K'$, both number fields, K′ a Galois extension of K with Galois group G, the respective rings of integers A and A′). Let $\mathfrak{p}$ be a maximal ideal of A which does not ramify in A′, and let $\mathfrak{p}'$ be a prime factor of $A'\mathfrak{p}$. The inertia group of $\mathfrak{p}'$ consists of the identity of G alone (§ 2, corollary of Proposition 2); its decomposition group D is canonically isomorphic to the Galois group of $A'/\mathfrak{p}'$ over $A/\mathfrak{p}$ (§ 2, Proposition 2). But the Galois group of $A'/\mathfrak{p}'$ over $A/\mathfrak{p}$ is *cyclic* with a canonical generator $\bar{\sigma}: \bar{x} \mapsto \bar{x}^q$, where $q = \text{card}(A/\mathfrak{p})$ (§ 1, Example 3). Therefore, D itself is cyclic with a canonical generator σ defined by the relation $\sigma(x) \equiv x^q \pmod{\mathfrak{p}'}$ for any $x \in A'$. This generator is called the *Frobenius automorphism* of $\mathfrak{p}'$. We denote it $(\mathfrak{p}', K'/K)$.

For $\tau \in G$ we have, as in the remark at the end of § 2,

$$(1) \qquad (\tau(\mathfrak{p}'), K'/K) = \tau\cdot(\mathfrak{p}', K'/K)\cdot\tau^{-1}.$$

In particular, if K′ is an abelian extension, $(\mathfrak{p}', K'/K)$ depends only on the ideal $\mathfrak{p}$ of A. In this case we write $\left(\frac{K'/K}{\mathfrak{p}}\right)$.

Proposition 1. With the preceding hypotheses and notations let F *be an intermediate field* ($K \subset F \subset K'$) *and write f for the residual degree of* $\mathfrak{p}' \cap F$ *over* K. *Then*

(a) $(\mathfrak{p}', K'/F) = (\mathfrak{p}', K'/K)^f$;

(b) *if* F *is Galois over* K, *the restriction of* $(\mathfrak{p}', K'/K)$ *to* F *equals* $(\mathfrak{p}' \cap F, F/K)$.

Proof. (a) Put $\sigma = (\mathfrak{p}', K'/K)$. By definition $\sigma(\mathfrak{p}') = \mathfrak{p}'$ and $\sigma(x) \equiv x^q \pmod{\mathfrak{p}'}$ for every $x \in A'$ (here $q = \text{card}(A/\mathfrak{p})$). Thus, $\sigma^f(\mathfrak{p}') = \mathfrak{p}'$ and $\sigma^f(x) \equiv x^{q^f} \pmod{\mathfrak{p}'}$ for all $x \in A'$. By definition of f, q^f is the cardinality of the residual field $(A' \cap F)/(\mathfrak{p}' \cap F)$. Furthermore, the decomposition group of $\mathfrak{p}'$ over F is obviously a subgroup of the decomposition group D of $\mathfrak{p}'$ over K. It is of order

$$[A'/\mathfrak{p}' : (A' \cap F)/(\mathfrak{p}' \cap F)] = f^{-1}[A'/\mathfrak{p}' : A/\mathfrak{p}] = f^{-1}\,\text{card}(D),$$

by (2) of § 2. Since D is cyclic and generated by σ, the only subgroup of D of order $f^{-1}\,\text{card}(D)$ is generated by σ^f. This proves (a).

(b) Suppose F is Galois over K and write σ' for the restriction to F of σ (§ 1, Theorem 1, (b)). Since $\sigma(\mathfrak{p}') = \mathfrak{p}'$, it follows that $\sigma'(\mathfrak{p}' \cap F) = \mathfrak{p}' \cap F$ and σ' belongs to the decomposition group of $\mathfrak{p}' \cap F$ over K. Moreover, it is clear that $\sigma'(x) \equiv x^q \pmod{\mathfrak{p}' \cap F}$ for all $x \in A' \cap F$, with $q = \mathrm{card}(A/\mathfrak{p})$. Q.E.D.

6.4. An application to cyclotomic fields

We are going to utilize the theory which we have just developed to present a third proof of the irreducibility of the cyclotomic polynomial (cf. Chapter II, § 9, Theorem 1 and Chapter V, § 2, Example).

Theorem 1. Let z be a primitive complex nth root of unity. Then

(a) *No prime number which does not divide n ramifies in* $\mathbf{Q}[z]$;

(b) $\mathbf{Q}[z]$ *is an abelian extension of* $\mathbf{Q}$ *of degree* $\varphi(n)$ *and with Galois group isomorphic to* $(\mathbf{Z}/n\mathbf{Z})^*$.

Proof. (a) Let F(X) be the minimal polynomial of z over $\mathbf{Q}$ and let d be its degree (so $d = [\mathbf{Q}[z] : \mathbf{Q}]$). The polynomial F(X) divides $X^n - 1$; let $X^n - 1 = F(X)\,G(X)$. We know that $D(1, z, \ldots, z^{d-1}) = \pm N(F'(z))$ (Chapter II, § 7, (6)). From the relation $nX^{n-1} = F'(X)G(X) + F(X)\,G'(X)$, we may prove, by substitution, $nz^{n-1} = F'(z)G(z)$. Since z is a unit of $\mathbf{Q}[z]$, it is of norm ± 1. Taking norms, we may conclude that $N(F'(z))$ divides n^d. Finally, since z is an integer in $\mathbf{Q}[z]$, the absolute discriminant of $\mathbf{Q}[z]$ divides $D(1, z, \ldots, z^{d-1})$ and, therefore also, n^d. By Chapter V, § 3, Theorem 1, no prime number which does not divide n ramifies in $\mathbf{Q}[z]$. This proves (a).

(b) Recall (§ 1, Example 2) that $\mathbf{Q}[z]$ is an abelian extension of $\mathbf{Q}$ and that there is an injective homomorphism j of the Galois group G of $\mathbf{Q}[z]$ over $\mathbf{Q}$ into $(\mathbf{Z}/n\mathbf{Z})^*$. More precisely, the element $\sigma \in G$ raises all the nth roots of unity to the power $j(\sigma)$. Let p be a prime number which does not divide n. By (a) the Frobenius automorphism $\left(\frac{\mathbf{Q}[z]/\mathbf{Q}}{p}\right)$ is defined; denote it σ_p. Writing A for the ring of integers of $\mathbf{Q}[z]$ and $\mathfrak{p}$ for an arbitrary prime factor of Ap, we obtain, from the definition of the Frobenius automorphism, the relation $\sigma_p(x) \equiv x^p \pmod{\mathfrak{p}}$ for all $x \in A$. In particular, putting $j = j(\sigma_p)$, we have $z^j \equiv z^p \pmod{\mathfrak{p}}$. Recall that $\prod_{\substack{0 \le r \le n-1 \\ r \not\equiv p \pmod n}} (z^p - z^r) = P'(z^p) = nz^{p(n-1)}$, where $P(X) = X^n - 1 = \prod_{0 \le r \le n-1} (X - z^r)$. Since n is relatively prime to p, since $\mathfrak{p} \cap \mathbf{Z} = p\mathbf{Z}$, and since z is a unit in the integers of $\mathbf{Q}[z]$, we may conclude from the relation $P'(z^p) = nz^{p(n-1)}$ that $\prod_{\substack{0 \le r \le n-1 \\ r \not\equiv p \pmod n}} (z^p - z^r) \notin \mathfrak{p}$. The relation $z^j \equiv z^p \pmod{\mathfrak{p}}$ thus implies that j represents the residue class of p mod n. Hence $j(G)$ contains the residue classes mod n of all the prime numbers p which do not divide n. But this means that $j(G) = (\mathbf{Z}/n\mathbf{Z})^*$, which proves (*b*).

6.5. Another proof of the quadratic reciprocity law

Let q be an odd prime. Let K be the cyclotomic field generated by a primitive qth root of unity in $\mathbf{C}$. The Galois group G of K over $\mathbf{Q}$ is isomorphic to $\mathbf{F}_q^*$ (§ 4, Theorem 1, (b)); it is cyclic of even order $q - 1$. There is a unique subgroup H of index 2, which corresponds to the subgroup of squares $(\mathbf{F}_q^*)^2 \subset \mathbf{F}_q^*$. Thus K contains a unique quadratic subfield F (§ 1, Theorem 2, (b)). No prime number $p \neq q$ ramifies in F, for, if it did, it would ramify in K; and this would contradict Theorem 1, (a), of § 4. The calculation of the discriminant of a quadratic field (Chapter V, § 3, Example 1) implies that $F = \mathbf{Q}[\sqrt{q}]$, if $q \equiv 1 \bmod 4$, and $F = \mathbf{Q}[\sqrt{-q}]$, if $q \equiv 3 \bmod 4$. Putting $q^* = (-1)^{(q-1)/2}q$, we have, in either case, $F = \mathbf{Q}[\sqrt{q^*}]$.

Let p be a prime distinct from q. Write σ_p for the Frobenius automorphism $\left(\frac{K/\mathbf{Q}}{p}\right)$ (cf. § 4). Its restriction to F is $\left(\frac{F/\mathbf{Q}}{p}\right)$ (§ 3, Proposition 1, (b)). It is the identity if $[\sigma_p \in H]$, i.e. if the exponent $j(\sigma_p)$ = the residue class of p mod q (cf. § 4) is a square in $\mathbf{F}_q^*$; otherwise it is the non-trivial automorphism of F. In other words, identifying the Galois group G/H of F over $\mathbf{Q}$ with $\{+1, -1\}$, we have

$$\left(\frac{F/\mathbf{Q}}{p}\right) = \left(\frac{p}{q}\right) \tag{1}$$

by definition of the Legendre symbol $\left(\frac{p}{q}\right)$ (Chapter V, § 5).

On the other hand, the results concerning the decomposition of the prime p in $F = \mathbf{Q}[\sqrt{q^*}]$ (Chapter V, § 4) give further information regarding $\left(\frac{F/\mathbf{Q}}{p}\right)$. By definition it is the identity if p splits in F, and it is the non-trivial automorphism if p remains prime. Using Proposition 1 of § 4, Chapter V, we may conclude that, if p is odd,

$$\left(\frac{F/\mathbf{Q}}{p}\right) = \left(\frac{q^*}{p}\right). \tag{2}$$

Comparing (1) and (2) we obtain

$$\left(\frac{p}{q}\right) = \left(\frac{q^*}{p}\right) = \left(\frac{-1}{p}\right)^{(q-1)/2}\left(\frac{q}{p}\right).$$

But

$$\left(\frac{-1}{p}\right) = (-1)^{(p-1)/2},$$

by Euler's criterion (Chapter V, § 5, Proposition 1). Consequently,

$$\left(\frac{p}{q}\right) = (-1)^{(p-1)(q-1)/4}\left(\frac{q}{p}\right).$$

This completes our second proof of the quadratic reciprocity law (Chapter V, § 5, Theorem 1).

To take care of the case $p = 2$, recall that 2 splits in F if $q^* \equiv 1 \pmod 8$ and 2 remains prime if $q^* \equiv 5 \pmod 8$ (Chapter V, § 4, Proposition 1). However, $(-1)^{(q^2-1)/8} = (-1)^{(q^{*2}-1)/8}$, and this is 1 if $q^* \equiv 1 \pmod 8$ and -1 if $q^* \equiv 5 \pmod 8$. Thus,

$$\left(\frac{F/\mathbf{Q}}{2}\right) = (-1)^{(q^2-1)/8}. \tag{3}$$

Putting (1) and (3) together, we obtain $\left(\frac{2}{q}\right) = (-1)^{(q^2-1)/8}$. This is the difficult "complementary formula" (Chapter V, § 5, Proposition 2, (b)).

A supplement, without proofs

This appendix presents, without proofs, some definitions and facts which complete and generalize certain ideas discussed in the text. The notions to be introduced here are not much deeper or more difficult than the corresponding material in the text. The author has kept a detailed discussion of these matters out of the text in order to keep the book short. Besides, the reader may find these more general results in numerous reference sources (see, for example, Chapter V of [9], which may be read immediately after this book).

It has not been the author's intention to summarize here, without proofs, developments in number theory which lie beyond the scope of this book (adèles, class fields, zeta functions and L-series, the arithmetic of simple algebras, analytic number theory, quadratic forms, etc.). He has preferred rather to refer the reader to the works cited in the first part of the bibliography (the unnumbered titles).

Transitivity formulas

Given three fields, $K \subset L \subset M$, each extension of *finite degree* over the preceding, we have "trace" mappings

$$\mathrm{Tr}_{L/K}: L \to K, \qquad \mathrm{Tr}_{M/L}: M \to L, \qquad \mathrm{Tr}_{M/K}: M \to K,$$

and, analogously, "norm" mappings (Chapter II, § 6). For any $x \in M$, we have the following relations

$$\text{(1)} \qquad \begin{cases} \mathrm{Tr}_{M/K}(x) = \mathrm{Tr}_{L/K}(\mathrm{Tr}_{M/L}(x)) \\ \mathrm{N}_{M/K}(x) = \mathrm{N}_{L/K}(\mathrm{N}_{M/L}(x)). \end{cases}$$

The relative norm of an ideal

Given two number fields, $K \subset K'$, and an ideal (integral or fractional) $\mathfrak{a}'$ of K', the ideal of K generated by the elements of K of the form $\mathrm{N}_{K'/K}(x)$ ($x \in \mathfrak{a}'$) is called the *relative norm* of $\mathfrak{a}'$. It is denoted $\mathrm{N}_{K'/K}(\mathfrak{a}')$ (or $\mathrm{N}(\mathfrak{a}')$). If $\mathfrak{a}'$ is a principal ideal (a'), then

$$\text{(2)} \qquad \mathrm{N}_{K'/K}(\mathfrak{a}') = (\mathrm{N}_{K'/K}(a')).$$

If $K = \mathbf{Q}$, this definition of norm coincides with that presented in Chapter III, § 5: if $\mathfrak{a}'$ is an integral ideal of K′ and if A′ is the ring of integers of K′ then

$$(3) \qquad N_{K'/\mathbf{Q}}(\mathfrak{a}') = \mathrm{card}(A'/\mathfrak{a}')\mathbf{Z}.$$

Let us return to the general case. If $\mathfrak{a}$ is an ideal of K and if $n = [K' : K]$, then

$$(4) \qquad N_{K'/K}(A'\mathfrak{a}) = \mathfrak{a}^n.$$

For $\mathfrak{a}'$ and $\mathfrak{b}'$ ideals of K′ there is the multiplicativity formula:

$$(5) \qquad N_{K'/K}(\mathfrak{a}'\,\mathfrak{b}') = N_{K'/K}(\mathfrak{a}')N_{K'/K}(\mathfrak{b}').$$

Finally, if $\mathfrak{p}'$ is a prime ideal of K′, if $\mathfrak{p} = \mathfrak{p}' \cap K$, and if f is the residual degree of $\mathfrak{p}'$ over K, then

$$(6) \qquad N_{K'/K}(\mathfrak{p}') = \mathfrak{p}^f.$$

For three number fields $K \subset K' \subset K''$ there is the following transitivity formula, in which $\mathfrak{a}''$ denotes an ideal of K″:

$$(7) \qquad N_{K''/K}(\mathfrak{a}'') = N_{K'/K}(N_{K''/K'}(\mathfrak{a}'')).$$

In the same context, in terms of the relative norm of an ideal one may obtain a transitivity formula for discriminants (where $\mathfrak{D}_{K'/K}$ denotes the discriminant of K′ over K; cf. Chapter V, § 3, Definition 1).

$$(8) \qquad \mathfrak{D}_{K''/K} = N_{K'/K}(\mathfrak{D}_{K''/K'}) \cdot \mathfrak{D}_{K'/K}^{[K'':K']}.$$

The above generalizes to the case of a Dedekind ring A and its integral closure A′ in a finite extension of the field of fractions of A.

The different

The following is valid for a Dedekind ring A and its integral closure in a finite extension of its field of fractions. For simplicity we limit our discussion to the number field case.

Let K and K′ be fields, $K \subset K'$. Let A and A′ be the respective rings of integers in K and K′. One says that a maximal ideal $\mathfrak{p}'$ of A′ is *ramified* over A (or over K) if its index of ramification over A is larger than 1. In this case one says that the maximal ideal $\mathfrak{p} = \mathfrak{p}' \cap A$ of A *ramifies* in A′ (Chapter V, § 3). It follows easily from Chapter V, § 3, Theorem 1 that the maximal ideals of A′ which are ramified over A are *finite in number*. We are going to describe an ideal $\mathfrak{d}_{K'/K}$ of A′, called the "different" of K′ over K, which will have the property that the maximal ideals ramified over A are precisely those which contain $\mathfrak{d}_{K'/K}$ (note the analogy with Chapter V, § 3, Theorem 1).

First, one proves that the set of all $x \in K'$ such that

$$(9) \qquad \mathrm{Tr}_{K'/K}(xA') \subset A$$

is a fractional ideal $\mathfrak{C}$ of A′; it is called the *codifferent* of K′ over K. By definition the *different* $\mathfrak{d}_{K'/K}$ is the inverse ideal $\mathfrak{C}^{-1}$. The different is a non-zero *integral* ideal of A′.

It can be shown that the different is generated by elements of the form $F'(x)$, where x runs through A' and where F denotes the minimal polynomial of x over K and where F' is the derivative of F. In particular, if A' is of the form $A[y]$ (which isn't always the case) and if G is the minimal polynomial for y over K, then the different $\mathfrak{d}_{K'/K}$ is the principal ideal of A' generated by $G'(y)$.

The non-zero prime ideals of A' which are ramified over A are those which contain $\mathfrak{d}_{K'/K}$. More precisely, let

$$\mathfrak{d}_{K'/K} = \prod_i \mathfrak{p}_i'^{m_i} \quad (m_i > 0) \tag{10}$$

be the decomposition of the different into a product of powers of prime ideals, and let e_i be the index of ramification of $\mathfrak{p}_i'$ over A. Then the non-zero prime ideals of A' which are ramified over A are the $\mathfrak{p}_i'$ and $m_i \geq e_i - 1$ for all i. Furthermore, $m_i = e_i - 1$ if and only if e_i is prime to the characteristic of the residual field $A'/\mathfrak{p}_i'$.

The different $\mathfrak{d}_{K'/K}$ (ideal of A') and the discriminant $\mathfrak{D}_{K'/K}$ (ideal of A) are connected by the following relation:

$$\mathfrak{D}_{K'/K} = N_{K'/K}(\mathfrak{d}_{K'/K}) \tag{11}$$

(cf. Chapter II, § 7, formula (6)). Thus the different gives more precise information concerning the ramification of primes in A and A' than does the discriminant.

Finally, for three number fields, $K \subset K' \subset K''$, there is the following transitivity relation for the differents:

$$\mathfrak{d}_{K''/K} = \mathfrak{d}_{K''/K'}\, \mathfrak{d}_{K'/K}. \tag{12}$$

Exercises

The exercises marked A are easy exercises which the reader may use to check his comprehension of the text. Those marked B are more elaborate. The "review problems" at the end have been given as examination problems in Paris.

Chapter I

1B. Let p be a prime number and let $r \in \mathbf{N}$. Show that the multiplicative group $(\mathbf{Z}/p^r\mathbf{Z})^*$ is cyclic, except if $p = 2$ and $r \geq 3$ (for p odd, show that the residue class of $1 + p$ is of order p^{r-1}. For $p = 2$ look at the order of the residue class of 5; then use Corollary 4 of Theorem 1, § 5).

2B. (a) Show that the diophantine equation $X^2 + Y^2 = 3Z^2$ has no non-trivial solution (i.e., $\neq (0, 0, 0)$) (Reduce mod. 3). Show the same for $X^2 + Y^2 = 7Z^2$ and $X^2 + Y^2 = 11Z^2$.

(b) Show that $X^2 + Y^2 = 5Z^2$ and $X^2 + Y^2 = 13Z^2$ have non-trivial integer solutions.

(c) Try to generalize (a) to certain equations $X^2 + Y^2 = pZ^2$, p a given prime (study under what condition -1 is a square in $\mathbf{F}_p^*$).

3B. Rediscover the classical proof for the fact that there are infinitely many prime numbers. By analogous methods show that there are infinitely many primes of the form $4k - 1$ $(k \in \mathbf{N})$.

4B. In order that $n \in \mathbf{N}$ be prime it is necessary and sufficient that n divide $(n - 1)! + 1$. (If n is prime, calculate the product of the elements of $\mathbf{F}_n^*$; check what happens in the case when n is not prime.)

5B. Show that in a finite field K every element is a sum of two squares (first treat the case where $q = \text{card}(K)$ is even; if q is odd, calculate the number of values taken by the functions $x \mapsto x^2$ and $y \mapsto a - y^2$, $x, y \in K$, $a \in K$ given).

6B. Factor the polynomial $X^3 - X + 1$ in $\mathbf{F}_{23}$ and the polynomial $X^3 + X + 1$ in $\mathbf{F}_{31}$ (each has a double root and a simple root).

7A. Give an example of two ideals $\mathfrak{a}$, $\mathfrak{b}$ of a ring A such that $\mathfrak{a} \cap \mathfrak{b} \neq \mathfrak{a} \cdot \mathfrak{b}$. Show that it is always the case that $\mathfrak{a} \cdot \mathfrak{b} \subset \mathfrak{a} \cap \mathfrak{b}$.

8B. Let A be an integral domain, a and $b \in A$, and let $B = A[X]/(aX + b)$. Show that, if $Aa \cap Ab = Aab$, then B is an integral domain (consider the element $-b/a$ of the field of fractions K of A, and show that the homomorphism $\varphi: A[X] \to K$ defined by $\varphi(X) = -b/a$ and $\varphi(y) = y$ for $y \in A$ has precisely $(aX + b)$ as its kernel).

9B. Let A be an integral domain and let i and j be relatively prime positive integers. Show that the ideal $(X^i - Y^j)$ in the polynomial ring $A[X, Y]$ is prime (define a homomorphism $A[X, Y] \to A[X]$ having that ideal as its kernel).

10B. (a) Let A be an integral domain, K its field of fractions, and b a non-zero element of A. Show that the A-module $\mathrm{Hom}(A/Ab, K/A)$ is isomorphic to A/Ab (recall that, given two A-modules E and F, we write $\mathrm{Hom}(E, F)$ for the set of homomorphisms, or A-linear maps, from E to F; recall that $\mathrm{Hom}(E, F)$ is an A-module. We note that a homomorphism $\varphi: A/Ab \to K/A$ is determined by its value at the residue class of 1).

(b) For an A-module, E, put $E' = \mathrm{Hom}(E, K/A)$ and put $E'' = \mathrm{Hom}(E', K/A)$. Define a canonical homomorphism $E \to E''$.

(c) Suppose that A is a principal ideal ring. Let E be a torsion module of finite type over A (i.e. such that, for every $x \in E$, there exists non-zero $a \in A$ for which $ax = 0$). Determine the structure of E and deduce from (a) that E is isomorphic to $E' = \mathrm{Hom}(E, K/A)$. Deduce from this that the canonical homomorphism $E \to E''$ is an isomorphism.

Chapter II

1A. Theorem 1 of § 7 remains true if, instead of supposing that K is of characteristic zero, we suppose K finite. Tell why. Is the case where K is finite of any interest?

2B. Let A be an integrally closed ring, K its field of fractions, and $P(X) \in A[X]$ a monic polynomial. If $P(X)$ is reducible in $K[X]$, show that it is reducible in $A[X]$ (consider the roots of $P(X)$ in an extension of K).

3B. For every positive integer n write $f(n)$ for the number of monic irreducible polynomials of degree n over $\mathbf{F}_q$. Show that $\sum_{d|n} df(d) = q^n$ (classify the elements of $\mathbf{F}_{q^n}$ according to their minimal polynomials over $\mathbf{F}_q$, and note that $\mathbf{F}_{q^d} \subset \mathbf{F}_{q^n}$ if and only if $d \mid n$). Deduce from this the values of $f(1), f(2), f(3), f(4)$, and $f(p)$ for p prime.

4A. Give an example of an integral domain which is not integrally closed (look at § 5).

5B. Let R be a ring, A a subring of R, and x a unit in R. Show that every $(y \in A[x] \cap A[x^{-1}]$ is integral over A (show that there exists an integer n such that the A-module $M = A + Ax + \cdots + Ax^n$ is stable under multiplication by y, i.e. that $yM \subset M$. One will at this point find the idea needed to conclude the proof by looking at the last part of the proof of Theorem 1, § 1).

Chapter III

1A. Why is a field a Dedekind ring?

2A. How could one modify the statement of Theorem 2, § 4 so that the hypothesis that A is not a field would become superfluous?

3B. In a Noetherian ring A it is clear that any ideal different from A itself is contained in a maximal ideal. Show that this is still true when A is not Noetherian (use Zorn's lemma).

4A. Give an example of an integral domain A containing a prime ideal $\mathfrak{p} \neq 0$ and a subring A′ such that $\mathfrak{p} \cap A' = (0)$.

5A. Let K be a field. Is the ring K[X, Y] of polynomials in two variables over K a Dedekind ring?

6. Let K be a field and let A be the subring $K[X^2, X^3]$ of the ring of polynomials K[X].

(a) By using the inclusion relations $K[X^2] \subset A \subset K[X]$, show that A is Noetherian and that every non-zero prime ideal is maximal.

(b) Show that A is not a Dedekind ring (consider the element X).

Chapter IV

1B. Show that the ring of integers A of the cubic field $\mathbf{Q}[x]$ with $x^3 = 2$ is principal (majorize the discriminant of A by $D(1, x, x^2)$ and apply Corollary 1 of Proposition 1, § 3; in fact, $A = \mathbf{Z}[x]$. See a review problem below).

2B. Determine the units of the ring $A = \mathbf{Z}[X]/(X^3)$ and the structure of the group A*.

3B. Let K be a cubic field such that $r_1 = r_2 = 1$. Suppose K is imbedded in **R**.

(a) Show that the positive units of K form a group isomorphic to **Z**, and that every positive unit of K is of norm 1.

(b) Let d be the absolute discriminant of K and let u be a unit greater than 1. Show that $|d| \leq 4u^3 + 24$ (put $u = x^2$ with $x \in \mathbf{R}$ and $x > 0$; note that the conjugates of u are of the form $x^{-1}e^{iy}$ and $x^{-1}e^{-iy}$ with $y \in \mathbf{R}$. Calculate the discriminant $d' = D(1, u, u^2)$ as a function of x and y, say $|d'|^{1/2} = \varphi(x, y)$. Find the minimum of $\varphi(x, y)$ for fixed x and deduce that $|d'| \leq 4u^3 + 24$. Conclude by observing that d divides d').

(c) Show that the polynomial $X^3 + 10X + 1$ is irreducible over **Q** (cf. Chapter V, § 3, example near the end). Let x be one of its real roots. Show that the ring of integers of the cubic field $\mathbf{Q}[x]$ is $\mathbf{Z}[x]$ (ibid.). Show that $u = -1/x$ is a generator of the group of positive units of $\mathbf{Q}[x]$.

4B. Let K be a number field, let A be the ring of integers of K, and let P denote the set of maximal ideals of A. For $\mathfrak{p} \in P$ and $x \in K^*$ write $v_\mathfrak{p}(x)$ for the exponent of $\mathfrak{p}$ in the factorization of Ax into a product of prime ideals. Put $v_\mathfrak{p}(0) = +\infty$.

(a) Show that, for $x, y \in K$, $v_\mathfrak{p}(xy) = v_\mathfrak{p}(x) + v_\mathfrak{p}(y)$ and $v_\mathfrak{p}(x + y) \geq \min(v_\mathfrak{p}(x), v_\mathfrak{p}(y))$.

(b) Let $P' \subset P$. Show that the set A(P′) of elements $x \in K$ such that $v_\mathfrak{p}(x) \geq 0$ for all $\mathfrak{p} \in P'$ is a subring of K. Consider the cases $P' = P$ and $P' = \phi$.

(c) Take for P′ the complement of a *finite* set $S \subset P$. Show that the group U of units of A(P′) is of finite type and that U/A* is a free **Z**-module of rank card(S) (consider the mapping $x \mapsto (v_\mathfrak{p}(x))_{\mathfrak{p} \in S}$ of U to $\mathbf{Z}^s$; find its kernel; determine its image by observing that every ideal of A has a power which is a principal ideal).

Chapter V

1B. Let A be a principal ideal ring and let K be its field of fractions.

(a) Show that every intermediate ring B $(A \subset B \subset K)$ is a ring of fractions of A (if $a/b \in \mathrm{B}$ with $a, b \in \mathrm{A}$ relatively prime, then show, by using Bezout's identity, that $1/b \in \mathrm{B}$).

(b) Classify the rings of fractions of A (consider the prime elements of A which are invertible in $\mathrm{S}^{-1}\mathrm{A}$).

(c) Show that any ring of fractions of A is principal.

2A. Show that no integer $\equiv 7$ modulo 8 is a sum of three squares (reduce mod 8).

3B. Let K be a field and A a reduced K-algebra of finite dimension over K.

(a) Making use of § 3, Lemma 3, show that A is isomorphic to a finite product $\prod_{i=1}^{n} \mathrm{L}_i$ of extensions of finite degree of K.

(b) Show that, if K is of characteristic 0, A has a primitive generator, i.e. A is of the form $\mathrm{K}[x]$ with $x \in \mathrm{A}$. (Find a set of primitive generators (x_i) for the fields (L_i) over K whose minimal polynomials over K are pairwise distinct. Take $x = (x_1, \ldots, x_n)$.)

4B. Let p be a prime number $\equiv 1$ mod 4. Show that there are exactly 8 pairs $(a, b) \in \mathbf{Z} \times \mathbf{Z}$ such that $a^2 + b^2 = p$ (in other words, the diophantine equation $p = a^2 + b^2$ has only one solution up to trivial transformations).

5B. Let K be a field, A an algebra of finite dimension over K. Give examples where A has no primitive generator (i.e. not of the form $\mathrm{K}[x]$):

(a) When A is not reduced;

(b) When A is reduced and K is finite (cf. Exercise 3).

6B. (a) Show that the polynomial $\mathrm{X}^5 - \mathrm{X} + 1$ is irreducible over $\mathbf{Q}$ (look at its reduction mod 5). Let x be one of its roots. Calculate the integers r_1 and r_2 for the field $\mathbf{Q}[x]$.

(b) Calculate the discriminant of $(1, x, \ldots, x^4)$ (over $\mathbf{Z}$; cf. Chapter II, § 7, formula (7)). Show that it is square-free and deduce that $\mathbf{Z}[x]$ is the ring of integers of $\mathbf{Q}[x]$ (§ 3, Proposition 1).

(c) Show that the ring $\mathbf{Z}[x]$ is principal (use Corollary 1 of Proposition 1 of § 3, Chapter IV and note, by reduction of $\mathrm{X}^5 - \mathrm{X} + 1$ mod 2 and mod 3, that $\mathbf{Z}[x]$ contains no ideals of norm 2 or 3).

7B. Let $\mathrm{K} = \mathbf{Q}[\sqrt{d}]$ be a quadratic field where d is a square-free integer and let A be the ring of integers of K.

(a) By means of Corollary 1, § 3, Chapter IV, show that A is principal for $d = 2$, $3, 5, 13, -1, -2, -3, -7$.

(b) By the same method show that any ideal of K is equivalent to an integer ideal of norm 1 or 2 for $d = 6, 7, 17, 21, 29, 33, -5, -11, -15, -19$. In which of the preceding fields does 2 remain prime ($d = 21, 29, -11, -19$; cf. § 4, Proposition 1)? Show that their rings of integers are principal.

(c) Suppose $d \equiv 2$ or 3 mod 4. Show that K contains a principal ideal of norm 2 if and only if there exist a and $b \in \mathbf{Z}$ such that $a^2 - db^2 = \pm 2$. Deduce that A is principal for $d = 6, 7$, and that it contains two ideal classes when $d = -5$ (note that 2 ramifies in this field).

(d) Suppose $d \equiv 1$ mod 4. Show that K contains a principal ideal of norm 2 if and only if $a^2 - db^2 = \pm 8$ has a solution $(a, b) \in \mathbf{Z} \times \mathbf{Z}$. Show that A is principal for $d = 17$,

33, and that it contains two ideal classes when $d = -15$ (note that 2 splits in this field).

(e) Show that, for $d = 10$ and $d = -6$, A contains two classes of ideals (note that any ideal of K is equivalent to an integral ideal of norm 1, 2, or 3; study the factorizations of 2 and 3 in K, and show that $(2 + \sqrt{10})(2 - \sqrt{10}) = -2 \cdot 3$, also $(\sqrt{-6})^2 = -2 \cdot 3)$.

(f) Show that, for $d = -23$, A contains three classes of ideals (Proceed as in (e). Show that there are non-principal prime ideals $\mathfrak{p}, \mathfrak{p}', \mathfrak{q}, \mathfrak{q}'$ of A with the property that $2A = \mathfrak{p}\mathfrak{p}'$ and $3A = \mathfrak{q}\mathfrak{q}'$. Observe that $x = \frac{1}{2}(3 + \sqrt{-23})$ and $y = \frac{1}{2}(1 + \sqrt{-23})$ are elements of A with norms 8 and 6, respectively. Study the factorization of Ax and Ay into prime ideals).

8A. Let K, K' and K" be number fields with $K \subset K' \subset K''$. Let $\mathfrak{p}''$ be a maximal ideal of K" and let $\mathfrak{p}' = \mathfrak{p}'' \cap K'$. Write e (respectively, e', e'') for the ramification index of $\mathfrak{p}''$ over K (respectively, $\mathfrak{p}'$ over K, $\mathfrak{p}''$ over K'). Write f (respectively, f' and f'') for the residual degree of $\mathfrak{p}''$ over K (respectively, $\mathfrak{p}'$ over K, $\mathfrak{p}''$ over K'). Show that $e = e'e''$ and $f = f'f''$.

9B. (a) Show that the polynomial $X^3 + X^2 - 2X + 8$ is irreducible over $\mathbf{Q}$ (note that it has no root in $\mathbf{Z}$, so no root in $\mathbf{Q}$ either). Write x for one of its roots, K for the cubic field $\mathbf{Q}[x]$, and A for the ring of integers of K.

(b) Show that $D(1, x, x^2) = 4 \cdot 503$ and 503 is prime.

(c) Show that $y = 4/x$ is an integer of K, that $y \notin \mathbf{Z}[x]$, and that $A' = \mathbf{Z} + \mathbf{Z}x + \mathbf{Z}y$ is a subring of K which contains $\mathbf{Z}[x]$ properly. Show that the discriminant of A' over $\mathbf{Z}$ is 503 (compare the discriminants of A and A', or calculate $D(1, x, y)$ directly). Conclude that $A = A'$.

(d) Show that 2A is the product of 3 distinct prime ideals (write the multiplication table of the elements $1, x, y$ and reduce mod 2). Deduce from this fact that there exists no element u of A such that $A = \mathbf{Z}[u]$ (note that A/2A is not primitively generated over $\mathbf{F}_2$).

(e) Show that, if u is a primitive element of K contained in A, the group $A/\mathbf{Z}[u]$ admits a quotient group of order 2 (note that two of the three homomorphisms of A on $\mathbf{F}_2$ given by (d) coincide on $\mathbf{Z}[u]$, and consider their difference). Conclude that $D(1, u, u^2)$ is a multiple of $4 \cdot 503$ (note that $D(1, u, u^2) = 503 \cdot \mathrm{card}((A/\mathbf{Z}[u])^2)$. Thus none of the numbers $D(1, u, u^2)$ generates the discriminant of K as an ideal in $\mathbf{Q}$.

(f) Show that the ring A is principal (by Chapter IV, § 3, Corollary 1 to Proposition 1 it suffices to show that every prime ideal of norm $n \leq 6$ is principal; exclude the case $n = 6$ and, by making use of (d), the case $n = 4$; then study the norms and the decomposition of $x, y, x - 1, x + 2$, and $x + 3$ in A; in order to settle the case $n = 2$, affirm that $(x - 2)/(x + 2)$ and $(x - 1)/(x + 3)$ generate two of the three prime ideals of norm 2; for $n = 3$, show that 3A is prime; to settle $n = 5$, show that 5A is the product of a prime ideal of norm 5 and of another of norm 25, and compare the factorizations and norms of 5A and of $(x + 1)A$).

10B. (a) Let p and q be distinct prime numbers ≥ 5 and such that $pq^2 \not\equiv 1 \pmod 9$. Let K be the cubic field $\mathbf{Q}[u]$ where $u^3 = p^2q$. Show that u and $v = pqu^{-1}$ are integers of K, that $D(1, u, v) = -27p^2q^2$, and that $B = \mathbf{Z} + \mathbf{Z}u + \mathbf{Z}v$ is a subring of the ring A of integers of K (note that $u^2 = qv$ and $v^2 = pu$).

(b) Let $\mathfrak{q}$ be a prime ideal of A containing Aq. Deduce from the fact that $u^3 = p^2q$ that the exponent of $\mathfrak{q}$ in the decomposition of Aq is 3, and that $Aq = \mathfrak{q}^3$. Show that $B \cap \mathfrak{q}$ is the ideal of B generated by u, v, q, that $A/\mathfrak{q} = B/B \cap \mathfrak{q}$, and that $A = B + \mathfrak{q}$. Conclude from this that $A = B + qA$. Similarly show that $Ap = \mathfrak{p}^3$ with $\mathfrak{p}$ prime and that $A = B + pA$.

(c) By considering $(u \pm 1)^3$ (which differs by one from the class of p^2q mod 3), show that 3A is the cube of a prime ideal of A, and that (as in (b)) $A = B + 3A$.

(d) Let $x = a + bu + cv$ ($a, b, c \in \mathbf{Q}$), an integer of K. Deduce, by calculating traces, that $3pqx \in B$, from which it follows that $3pqA \subset B$. Use (b) and (c) to show that $A = B + 3pqA$. Conclude from this that $A = B$.

(e) Show that, for $x = a + bu + cv$ ($a, b, c \in \mathbf{Z}$),

$$D(1, x, x^2) = 3p^2q^2(b^3q - c^3p)^2.$$

(f) Deduce from (e) that, if $p \equiv 1 \pmod 3$ and $q^{(p-1)/3} \equiv 1 \pmod p$, A has no base (over $\mathbf{Z}$) of the form $(1, x, x^2)$ ($x \in A$). Show that this is the case for $p = 7$ and $q = 5$ (in which case $K = \mathbf{Q}[\sqrt[3]{175}] = \mathbf{Q}[\sqrt[3]{245}]$). Show that, in this case, the absolute discriminant of K is the greatest common divisor of the numbers $D(1, x, x^2)$, for $x \in A$.

Chapter VI

1B. (a) Show that the field $L = \mathbf{Q}[\sqrt{5}, \sqrt{-1}]$ is a Galois extension of degree 4 of $\mathbf{Q}$. Determine its Galois group.

(b) Show that the ring A of integers of L is $\mathbf{Z}[\sqrt{-1}, (1 + \sqrt{5})/2]$ (note that the discriminant of $(1, (1 + \sqrt{5})/2)$ over the ring of Gaussian integers is square-free (i.e. no Gaussian integer square factors); then make use of ideas from Chapter V, § 3, Proposition 1). Calculate the absolute discriminant of L. Show that the only primes which ramify in L are 2 and 5, and that the corresponding ramification indices are both 2 (use the fact that $n = efg$, cf. § 2, Proposition 1).

(c) Calculate the Frobenius automorphisms $\left(\frac{L/\mathbf{Q}}{p}\right)$ for p prime distinct from 2 and 5 (cf. § 3). Calculate the decomposition and inertia groups for 2 and 5.

(d) Show that $\mathbf{Q}[\sqrt{-5}] \subset L$, and that no prime ideal of $\mathbf{Q}[\sqrt{-5}]$ ramifies in L.

(e) Similarly, show that $\mathbf{Q}[\sqrt{10}] \subset \mathbf{Q}[\sqrt{2}, \sqrt{5}]$, and that no prime ideal of $\mathbf{Q}[\sqrt{10}]$ ramifies in $\mathbf{Q}[\sqrt{2}, \sqrt{5}]$.

Note that both $K = \mathbf{Q}[\sqrt{-5}]$ and $K = \mathbf{Q}[\sqrt{10}]$, by the above, admit abelian "unramified" extensions (i.e. in which no prime ideal of K ramifies) whose Galois group (cyclic of order 2) is isomorphic to the group of ideal classes of K. This is a very special case of the celebrated "class-field theory".

2B. Let x be the real root of the polynomial $X^3 - X + 1$, y and $\bar{y}$ its two other roots in $\mathbf{C}$, and K the cubic field $\mathbf{Q}[x]$ (cf. Chapter V, § 3, example).

(a) Show that $y + \bar{y} = -x$, $y\bar{y} = -1/x$, and $[(y - x)(\bar{y} - x)(y - \bar{y})]^2 = -23$ (cf. Chapter II, § 7, formulas (6) and (7)). Conclude from this that $K[\sqrt{-23}] = \mathbf{Q}[x, y, \bar{y}]$ and that this field L is a Galois extension of degree 6 of $\mathbf{Q}$. Determine its Galois group G (note that a $\mathbf{Q}$-automorphism of L is determined by its effect on $x, y, \bar{y}$).

(b) What are the subgroups of G? Determine the corresponding subfields of L (cf. § 1, Theorem 2). What are the subfields of L which are Galois over **Q**?

(c) Recall that the ring of integers of K is $\mathbf{Z}[x]$ (Chapter V, the example at the end of § 3). Show that it is principal (Chapter IV, § 3, Corollary 1 of Proposition 1).

(d) Show that 23 is the only prime number ramified in L, and that the corresponding ramification index is 2 (note that the decomposition of 23 into prime factors in the ring of integers of K is of the form $\alpha\beta^2$. Denoting the numbers defined in Proposition 1 of § 2 relative to the decomposition of 23 in L by e, f, and g, deduce that $g \geq 2$ and that e is even (cf. exercise 8 of Chapter V). Use the fact that $efg = 6$ to conclude that $e = 2, f = 1, g = 3$).

(e) Show that $\mathbf{Q}[\sqrt{-23}] \subset$ L and that no prime ideal of $\mathbf{Q}[\sqrt{-23}]$ ramifies in L.

Observe that L is a cyclic unramified extension of degree 3 of $\mathbf{Q}[\sqrt{-23}]$, and that the ideal class group of $\mathbf{Q}[\sqrt{-23}]$ is cyclic of order 3 (cf. Exercise 7, (f) of Chapter V). This is another special case of class field theory.

3B. Reformulate exercise 2 to apply to the polynomial $X^3 + X + 1$ (instead of $X^3 - X + 1$) and to -31 (in place of -23). Reformulate for $\mathbf{Q}[\sqrt{-31}]$ Exercise 7, (f) of Chapter V.

4B. Let K be a number field, and let L be the logarithmic imbedding of K^* in $\mathbf{R}^{r_1+r_2}$ (Chapter IV, § 4, Theorem 1). We write U for the group of units of K and we call the volume of the lattice L(U) of the hyperplane W considered in Chapter IV, § 4, Theorem 1 (with respect to the measure on W induced from Lebesgue measure by the mapping $W \to \mathbf{R}^{r_1+r_2-1}$) the *regulator* of K.

(a) Calculate the regulators of the real quadratic fields given as explicit examples in Chapter IV, § 6.

(b) Let z be a primitive nth root of unity in **C**, let K be the cyclotomic field $\mathbf{Q}[z]$ let σ be the automorphism of K defined by $\sigma(z) = z^{-1}$ and K′ the field left fixed by σ. Calculate $[K : K']$ and the ranks of the groups of units U and U′ of K and K′, respectively. Show that these ranks are equal to the same integer r. Show that there exists an integer m such that the regulators ρ and ρ' of K and K′ are connected by the relation $\rho' = m\rho$. Calculate the number of roots of unity contained in K (respectively, K′) (one may take $z = e^{2i\pi/n}$). Determine the index of U′ in U.

(c) With the notations of (b), suppose that $n = p^s$ with p prime, $p \neq 2$. Show that, for $u \in U$, we have $\sigma(u) = \pm z^{2j}u$ (note that $\sigma(u)u^{-1}$ has absolute value one for any imbedding of K in **C**, thus is a root of unity), whence $\sigma(z^j u) = \pm z^j u$. Show that $\sigma(z^j u) = \pm z^j u$. Show that $\sigma(z^j u) = z^j u$ (exclude the minus sign by considering the conjugates of $z^j u$ over K′). Deduce that $U = V \cdot U'$, where V is the group of nth roots of unity contained in K. Show that the integer m of (b) is equal to 1.

5B. (a) Let K be the cyclotomic field generated over **Q** by a ninth root of unity z. Show that it is a cyclic extension of degree 6 of **Q** (cf. § 4). Let G be its Galois group.

(b) By making use of the subgroup H of index 3 of G, construct a Galois extension K′ of degree 3 of **Q** (cf. § 1, Theorem 2). Show that $K' = \mathbf{Q}[\cos 2\pi/q]$ and determine the minimal polynomial of $2 \cos 2\pi/q$ over **Q** (it has as roots $z + z^8$, $z^2 + z^7$, and $z^4 + z^5$).

REVIEW PROBLEMS

I

In the field $\mathbf{C}$ of complex numbers, put $z = e^{2i\pi/5}$ and write K for the cyclotomic field $\mathbf{Q}[z]$.

1. Find the degree, the ring of integers, and the absolute discriminant of K (to calculate the absolute discriminant it will be convenient to calculate the numbers $\mathrm{Tr}(z^j)$). What primes ramify in A?

2. Write $\mathbf{R}$ for the field of real numbers and put $K' = K \cap \mathbf{R}$. Show that K′ is a quadratic field and determine which one. Describe the ring A′ of integers of K′ and the group A'^* of units of A′.

3. Show that K′ is the only quadratic field contained in K (note that, if K″ is a subfield of K, every prime number which ramifies in K″ ramifies in K).

4. Show that K contains no roots of unity besides the tenth roots of unity (it suffices to consider those roots of unity whose order is prime or a power of 2, and to make use of (3)). Write G for the group of tenth roots of unity.

5. (a) Making use of the logarithmic imbedding of K, show that, if a is a unit of absolute value one in the ring A of integers of K, then a is a root of unity.

(b) Let $a = re^{i\theta}$ be a unit of A ($r > 0$). Show that $r^2 \in A'^*$ and $e^{2i\theta} \in G$. Also show that the relations $r \in A'^*$ and $e^{i\theta} \in G$ are equivalent.

(c) Recall that $A(1-z) = A(1-z^2) = A(1-z^3) = A(1-z^4)$ is a prime ideal $\mathfrak{p}$ of A and that $5A = \mathfrak{p}^4$. Conclude that no element of $(1-z^3)K'$ is a unit of A, that A contains no unit of argument $\pi/10$, and that the group A^* of units of A is a direct product of G and of the group U of positive units of A′.

6. Recall that, with the classical notations, every class of ideals of a number field contains an integral ideal of norm $\leq (4/\pi)^{r_2} n!\, n^{-n}\, |d|^{1/2}$. Show that the ring A is principal.

II

Let $\mathbf{Q}$ be the field of rational numbers, x the real number such that $x^3 = 2$, and K the field $\mathbf{Q}(x)$.

1. Show that the polynomial $X^3 - 2$ is irreducible over $\mathbf{Q}$, and that K is an extension of degree 3 of $\mathbf{Q}$. In the field $\mathbf{C}$ of complex numbers what are the fields conjugate to K?

2. Let $z = a + bx + cx^2$ ($a, b, c \in \mathbf{Q}$) be an element of K. Calculate the trace, norm, and characteristic polynomial of z over $\mathbf{Q}$.

3. Let $\mathbf{Z}$ be the ring of rational integers and B the subring $\mathbf{Z}[x]$ of K. In this problem we intend to show that the ring A of integers of K is equal to B.

(a) Show that $B \subset A$, and find a base for B as a $\mathbf{Z}$-module.

(b) Let $z = a + bx + cx^2$ ($a, b, c \in \mathbf{Q}$) be an integer of K. By calculating the traces of z, xz, and x^2z show that $6A \subset B$.

For the rest of the proof one may, by making use of (2), show that the integers $6a$, $6b$, and $6c$ defined in (b) are multiples of 6 (the argument is long, but elementary). One may also proceed, in a less elementary way, as follows:

(c) Show that the ideals xA and xB are prime (for xA one may consider the decomposition of the ideal $2A$ in A). Show that $xB = B \cap xA$, and that the fields A/xA and B/xB are the same. Conclude that $A = B + xA$ and, hence, that $A = B + 2A$.

(d) Show that $3 = (x - 1)(x + 1)^3$ and that $x - 1$ is invertible in B. Proceed as in (c), with $x + 1$ in place of x and 3 in place of 2, to show that $A = B + (x + 1)A$, and consequently $A = B + 3A$.

(e) Deduce from (c) and (d) that $A = B + 6A$, and, finally, deduce that $A = B$.

4. Calculate the different and the discriminant of K.

5. (a) What is the structure of the (multiplicative) group of units of A? Show that every positive unit (respectively, negative unit) is of norm 1 (respectively, -1). Show that $x^2 + x + 1$ is a unit of A.

(b) Recall that, if u is a unit >1 of A and if D is the discriminant of K, then $|D| < 4u^3 + 24$ (cf. Exercise 3 of Chapter IV). Conclude that $x^2 + x + 1$ is a generator of the group of positive units of K (the approximate values of x and x^2 are 1.26 and 1.58).

(c) Calculate 3 or 4 solutions in integers a, b, c of the equation $a^3 + 2b^3 + 4c^3 - 6abc = 1$. Explain how one may find all its rational integer solutions, also all its positive integer solutions.

III

1. Show that the polynomial $X^3 - 3X + 1$ has no root in $\mathbf{Q}$ and is irreducible over $\mathbf{Q}$. Show that its three roots in $\mathbf{C}$ are real; let x be one of them.

Write K for the cubic field $\mathbf{Q}[x]$ and A for the ring of integers of K.

2. Calculate the discriminant $D(1, x, x^2)$. Deduce that, if $a + bx + cx^2$ $(a, b, c \in \mathbf{Q})$ is an integer of K, then $a, b, c \in S^{-1}\mathbf{Z}$ where S is the set of powers of 3, $\{3^n\}_{n \geq 0}$ (consider $S^{-1}\mathbf{Z} \subset S^{-1}A$).

3. Put $A' = \mathbf{Z}[x]$. Show that x and $x + 2$ are units of A' and of A, and that $(x + 1)^3 = 3x(x + 2)$. Conclude that $(x + 1)A$ is a prime ideal of A, that $(x + 1)A \cap A' = (x + 1)A'$, and that $A/(x + 1)A = A'/(x + 1)A'$. Show that $A = A' + (x + 1)A$ and $A = A' + 3A$. Conclude from this and from (2) that $A = A' = \mathbf{Z}[x]$.

4. Show that $2A$ is a prime ideal of A (note that $X^3 - 3X + 1$ is irreducible mod 2). Deduce from this that A is a principal ideal ring (recall that every ideal class of A contains an integral ideal of norm $\leq (4/\pi)^{r_2}(n!/n^n)\,|D|^{1/2}$, with the usual notations).

5. Show that $x^2 - 2$ is a root of the polynomial $X^3 - 3X + 1$. What are the fields conjugate to K over $\mathbf{Q}$, and what are the $\mathbf{Q}$-isomorphisms of K into $\mathbf{C}$?

6. Recall that $\cos 3u = 4\cos^3 u - 3\cos u$. Putting $x = 2\cos u$, calculate the possible values for the angle u. Conclude that K is of the form $L \cap \mathbf{R}$, where L is a cyclotomic field. Determine which cyclotomic field.

N.B. Questions 5 and 6 are independent of questions 2, 3, and 4.

Bibliography*

E. ARTIN, *Theory of algebraic numbers* (G. Striker, Schildweg 12, Göttingen, Germany—1957) (gives a presentation of valuation theory; very elegant; numerous examples).

H. HASSE, *Zahlentheorie* (Akademie Verlag, Berlin, 1949) (a large and very complete work).

H. HASSE, *Vorlesungen über Zahlen theorie* (Springer, 1964) (describes many aspects of number theory).

G. H. HARDY and E. M. WRIGHT, *An introduction to the theory of numbers* (Clarendon Press, Oxford, 1965) (profound and beautiful; shows a remarkable aesthetic sense in the choice of material).

E. HECKE, *Vorlesungen über die Theorie der algebraischen Zahlen* (Chelsea, New York, 1948) (a classic; elegant and complete).

S. LANG, *Algebraic numbers* (Addison-Wesley, 1964) (a little book—very dense and concentrated).

S. LANG, *Diophantine geometry* (Interscience Tract no. 11, J. Wiley, New York, 1962) (oriented toward diophantine equations; gives a nice indication of their connection with algebraic geometry).

O'MEARA, *Introduction to quadratic forms* (Springer, 1963) (an efficacious exposition of algebraic number theory, followed by one of its most beautiful applications and motivations).

J. P. SERRE, *Corps locaux* (Hermann, Paris, 1962) (the emphasis here is on p-adic fields; a very clear and lucid exposition of the most recent algebraic methods in number theory; very rich in content; numerous examples).

E. ARTIN and J. TATE, *Class-field theory* (W. A. Benjamin, 1968) (the most modern exposition of the famous class-field theory, i.e. the theory of abelian extensions of number fields).

A. WEIL, *Basic number theory* (Springer, 1967) (utilizes the fruitful method of adèles and treats simultaneously number fields and function fields).

Z. I. BOREVICH and I. R. SHAFAREVICH, *Number theory* (Academic Press, 1966) (very complete; excellent chapters on complex and p-adic analytic methods; numerous numerical tables).

* The references preceded by a number between brackets are referred to in the text.

[1] N. Bourbaki, *Algèbre* (Paris, Hermann) (especially Chapter V which discusses fields, Chapter VI which discusses divisibility, and Chapter VII which discusses modules over principal ideal rings).
[2] N. Bourbaki, *Algèbre commutative* (ibid.) (especially Chapter V, which deals with integrality, and Chapter VII, which discusses Dedekind and factorial rings; Chapter II gives a very complete and general discussion of rings of fractions and there is a good exposition of valuation theory in Chapter VI).
[3] C. Carathéodory, *Theory of functions*, Vol. I (Chelsea, New York, 1954).
[4] S. Lang, *Algebraic structures* (Addison-Wesley, 1967).
[5] S. Lang, "On quasi-algebraic closure" (*Ann. of Math.*, 55 (1962), 373–390).
[6] P. Samuel, "A propos du théorème des unités" (*Bull. Sci. Math.*, 90 (1966), 89–96).
[7] P. Samuel, "Anneaux factoriels" (*Publ. Soc. Math. São Paulo*, 1964) (also see the article: "Unique factorization," *American Math. Monthly*, 75 (1968), 945–952).
[8] G. Terjanian, "Sur une conjecture de M. Artin" (*C.R. Acad. Sci. Paris*, (1966)).
[9] O. Zariski and P. Samuel, *Commutative algebra*, Vol. I (Van Nostrand, Princeton, 1958) (Chapter II discusses field theory, Chapter IV discusses Noetherian rings, and Chapter V discusses integrality and Dedekind rings).

Index

The notation V.2 stands for Chapter V, § 2.

The index, I copied from old
Vladivostok telephone directory
Tom Lehrer